Dennett's Real Patterns in Science and Nature

Dennett's Real Patterns in Science and Nature

edited by Tyler Millhouse, Steve Petersen, and Don Ross

The MIT Press
Cambridge, Massachusetts
London, England

This book was set in Times New Roman by the authors. Printed and bound in the United States of America.

Library of Congress Cataloging-in-Publication Data

Names: Millhouse, Tyler editor | Petersen, Stephen editor | Ross, Don, 1962-editor
Title: Dennett's Real patterns in science and nature / edited by Tyler Millhouse, Stephen D. Petersen, and Don Ross.
Description: Cambridge, Massachusetts: The MIT Press, [2026] | Includes bibliographical references and index.
Identifiers: LCCN 2025024102 (print) | LCCN 2025024103 (ebook) | ISBN 9780262052030 paperback | ISBN 9780262052047 pdf | ISBN 9780262052054 epub
Subjects: LCSH: Dennett, D. C. (Daniel Clement)–Influence | Philosophy of mind | Philosophy of science
Classification: LCC B945.D394 D475 2026 (print) | LCC B945.D394 (ebook)
LC record available at https://lccn.loc.gov/2025024102
LC ebook record available at https://lccn.loc.gov/2025024103

EU Authorised Representative: Easy Access System Europe, Mustamäe tee 50, 10621 Tallinn, Estonia | Email: gpsr.requests@easproject.com

I walk down the garden paths,
And all the daffodils
Are blowing, and the bright blue squills.
I walk down the patterned garden paths
In my stiff, brocaded gown.
With my powdered hair and jewelled fan,
I too am a rare
Pattern. As I wander down
The garden paths.

—Amy Lowell, "Patterns"

It is interesting to contemplate a tangled bank, clothed with many plants of many kinds, with birds singing on the bushes, with various insects flitting about, and with worms crawling through the damp earth, and to reflect that these elaborately constructed forms, so different from each other, and dependent upon each other in so complex a manner, have all been produced by laws acting around us.

—Charles Darwin, *On the Origin of Species*

Contents

Dedication: Dan Dennett Is a Real Pattern

This book grew out of a workshop in which the most distinguished participant was Daniel C. Dennett. The volume was to have included his reflections on each of its chapters, a prospect that gave special motivation to all the authors—what commentator on science wouldn't itch to have their ideas refracted, however critically they might be received, through the mind of Dennett?

We were shocked to have these expectations dashed—because only one thing ever could, and finally actually did, keep Dan from fulfilling a commitment. The chapter drafts were delivered to him in mid-April of 2024. Uncharacteristically, no acknowledgment of receipt quickly flew back. A few days later we found out why.

It is trite to say that a person's death is the demise of a bundle of integrated memories, but perhaps a bit less so when what has vanished is a set of capacities for insight so unique that thousands of other thoughtful minds eagerly attended to all its products that were made available to them. It is a special jolt from forgetfulness, the ultimate system crash, to have such an irreplaceable resource erased at a stroke. Several times during the first days after the news came in, I found myself wondering, "Now how are we going to keep our thoughts in order?"

As the shock abated, the answer to that question I'd asked myself started to settle in. Of course, we really will think a bit less clearly than we used to, because a powerful critical filter is gone. But a great many people will go on thinking better with Dan's help, because we can still interact with him virtually: We can still simulate, when we consider a new topic or proposition, what Dan would probably have thought about it. But how can such exercises actually be useful? Me "simulating" hypothetical thoughts of Dan's is in fact me thinking using my own cognitive resources, not magically conjuring his. My own apparent insights are not improved in quality by imagining them expressed in Dan's voice. But here is a nearby idea that might be true: It is evidence for the value of a line of thought if I can genuinely conceive of it in Dan's style.

Or, well, maybe; this still looks at first glance like complacent hocus-pocus. But it was Dan himself, in his work on consciousness and the social externality of mind, who showed us why there might be a real mechanism behind the trick of "thinking through Dan." When a person deliberates by silently framing her thoughts in her favored natural language, Dan argued in *Consciousness Explained*, she thereby forces herself to consult the accumulated wisdom of her culture, because she must think in forms that someone else from that culture could understand. If there are no words for your thought—if, for example, it is about "ineffable" qualia—then it is probably an unreliable thought. On the other hand, if you can't avoid clichés in expressing it, then it is just something that everyone knows already. To think both clearly and originally, you have to engage in dialogue with your intellectual ancestors.

If the simulated person you talk with when you address yourself in English or French or isiXhosa is no one in particular, but just a representative cultural agent, then you have just done a minimal quality check. To go to the next level of value, you must engage with a simulated mind that is more consistent and focused in its commitments than that generic representative. And you can only do that with a mind you've known very well. Given that condition, you can mine real value. Can a physicist do better physics by occasionally consulting with a dedicated biographer of Bohr? Yes—that expert, to a nontrivial extent, knows how to think like Bohr.

A mind, Dan taught us, is a real pattern. It is a pattern because it has become coherent, in a specific way, under the pressure of interacting with many people along increasingly well-explored tracks, so as to be a recognizable element in the world. It is a real pattern because it is projectable to new circumstances. The point of building a rich self, the reason your self is so important that you might rather be killed than betray it too egregiously, is that it allows others to make informed conjectures about what you'll do and say under various contingencies. In the normal run of things, you want a self that some others—those who "know" you—will co-invest with, and that a few others might love. That is all there is to being that kind of real pattern, a person, and to creatures like us with dense batteries of frontal cortical neurons, which can support imaginative construction and projection, nothing is more important.

Having a special *intellectual* self doesn't matter much to most people (and there's no reason it should), but it must tightly govern the self-construction of the minority who hope to guide others to better understand the world. Of course Dan was an exemplary member of this minority, and he built an unusually capacious and carefully mapped self that others could use to help generate good ideas and undermine bad ones. The projectability of Dan's mind as a real pattern was always crucial, since that is what made him a great philosopher rather than just someone who knew a lot about cognition and its evolution. (That is, we could see how new things he said lent support to, and were supported by, other things he'd said, across a wide breadth of topics.)

Since April 2024, that rigorously built projectability, where Dan is concerned, has acquired much greater value on the margin, because he can no longer actively refine it. It is as projectable as it ever was, so long as we don't let ourselves forget to attend to it (a grave new risk, since Dan will no longer prod us with new books and emails). A distinct, projectable generator of fresh consequences is all that a real pattern is, and to be is to be a real pattern. Dan's brain and consciousness are no more, but the pattern that is his mind is as real as it ever was.

To the extent that someone knows that pattern carefully enough to project it with better-than-chance success, they can still use the real pattern of Dan's mind to try to keep thinking straight.

Of course there was more to Dan's self than his intellectual ideas; he was loved as well for his generosity, his sense of justice, and his consoling wit. Those aspects of the real pattern of his mind will, alas, fade away when those who interacted with more than his ideas follow him in losing our own consciousnesses once and for all. And no real pattern—including the human species, the earthly Darwinian web of life, and perhaps even the universe as a whole—literally lasts forever. Furthermore, real patterns are all dynamical systems, so they are not immutable even while they persist. So Dan's mind is not immortal, and he would have appreciated my saying so. But it is far from gone now, and we help ourselves by taking good care to keep this precious resource close at hand.

The chapters in this book all represent efforts to do that.

—Don Ross

Acknowledgments

We would first like to thank Philip Laughlin of the MIT Press for his support and guidance throughout the publication process. We would also like to extend a special thanks to Nelleke Bak for her invaluable editorial assistance in preparing this volume. We also want to acknowledge and thank all those who aided in reviewing and editing individual chapters, including Valia Allori, Helen Beebee, Ray Jackendoff, Harold Kincaid, David Kinney, and Amanda Millhouse. In addition, we thank the Santa Fe Institute for hosting the workshop on which this volume is based and the National Science Foundation for its support of that workshop (Grant #2020103). We also thank *The Journal of Philosophy* for permission to reprint "Real Patterns," by Daniel C. Dennett (vol. 88, pp. 27–51, 1991). Finally, warm thanks to Susan Dennett for helping us find a suitable picture of Dan in his autumnal period for the cover.

1 An Introduction to Real Patterns

Tyler Millhouse, Steve Petersen, and Don Ross

1.1 An Intellectual Project

Our world is a richly structured tapestry. It is home to objects with their properties and relations, dynamical systems with their states and evolution rules, species with their genotypes and phenotypes, and agents with their beliefs and desires. It also is also home to a promiscuous assortment of more specific entities, including cabbages and kings, elms and beeches, Moses and Nixon, white swans and black ravens, desert landscapes and crowded rain forests—enough to make the most ardent stamp collector blush.

Many of these are part of what Sellars (1962) has called the "manifest image," the conceptual framework by which we first understand the world and our place within it (41). Some have imagined that science will pare down this bloated commonsense ontology, replacing some or all of it with a more parsimonious picture of the world. Certainly, the manifest image *has* changed in response to scientific advancement (as Sellars notes), but these changes rarely simplify our commonsense ontology.

The cosmology of a contemporary layperson, for example, would be unrecognizable to a human living in ancient Mesopotamia, but this is not because astronomy has aggressively pared down ancient cosmology. Even the tiny fraction of modern cosmology captured by contemporary commonsense is vastly more elaborate than the parochial cosmology of the ancient Near East. This ancient cosmology envisioned a universe centered on a small but important area of the world. Modern cosmology envisions the Earth lost in a sea of stars, planets, nebulae, and the like. These entities can be further sorted into any number of categories that illuminate their underlying physics. Stars, for example, are categorized by the clusters and bands of a Hertzsprung–Russell diagram (Hertzsprung 1911; Russell 1913). Stars also participate in larger patterns and structures, whether quotidian open clusters or stunning flocculent spirals.

In this regard, cosmology is unexceptional. The "scientific image" routinely exhibits a greater ontological promiscuity than the manifest image. Still, there is more to the distinction than this (Sellars 1962, 41). From the tree of life to the eightfold way, science has attempted to organize its wide-ranging ontology with an eye toward empirical results, conceptual clarity, and scientific explanation. These efforts are inextricably linked to the growth of scientific knowledge. The periodic organization of the chemical elements, for example, began as an attempt to regiment observed differences between substances. Further inquiry into these differences led to their explanation in terms of repeating patterns in the configurations of electrons, decades after Mendeleev's pioneering work on the periodic table as a classification scheme (Bohr 1913, 1923; Mendeleev 1869).

The popular image of science as debunking elements of folk belief seems to arise from the many cases in which science has disabused us of quaint superstitions (e.g., the existence of witches or the influence of heavenly bodies on human lives). Science has also promoted its reputation for parsimony by endorsing fewer kinds of substance (e.g., by abandoning dualism in favor of materialism about the mind). Even so, the progress of science has hardly spared its own ontology. New discoveries have made it untenable to affirm many things that were previously members in good standing of the scientific image of the world (e.g., the four humors). As with cosmology, the replacements for these discarded entities could hardly be viewed as simpler (e.g., modern biochemistry).

Even if we set aside the idea that science will deliver an ontological "desert landscape" (Quine 1948), we might still hope to replace our commonsense ontology with a scientific one. Here we encounter significant difficulties as well. First, many issues of deep personal and ethical concern are bound up with elements of the manifest image (e.g., persons, free will, or beauty). Second, it is not clear where to draw the line between the scientific and manifest images. Science has borrowed and refined many elements of the manifest image, including specific entities (e.g., the seven classical planets), general phenomena (e.g., heat), and conceptual schemes for knowledge representation (e.g., sounds, numbers, and shapes). Further, the very methodology of science is arguably an extension of commonsense, with science being distinguished only by the care with which commonsense is applied (Quine 1957). This suggests a continuity between the images that belies simplistic adversarial narratives.

Thus, scientists and philosophers are left with an important project: to understand how the elements of *both images* "hang together in the broadest possible sense"—that is, with "no intellectual holds . . . barred" (Sellars 1962, 37). As part of this project, elements of scientific or commonsense ontology that appear to resist this unification may be discarded, revised, or (despite first appearances) unified by subtler philosophical or scientific explanations. Mercifully, this daunting project can be subdivided into a host of more specific scientific and philosophical questions. In the section that follows, we will introduce one such question and Dennett's (1989, 1991) approach to answering it.

1.2 A Mild Realism

1.2.1 Facing Up to Eliminativism

As noted earlier, the story of both images has been one of continual revision. These revisions, along with other epistemic and ontological considerations, have caused more than a few to entertain various forms of irrealism. Some have focused their ontological ire on elements of the manifest image, such as the beliefs and desires that figure in folk psychology (see, e.g., Churchland 1981), while others have defended more general forms of scientific irrealism (see, e.g., van Fraassen 1980). Whatever their other merits, these irrealisms have been dogged by the stubborn utility of both images. Whether bosons or beliefs, our postulates give us startling predictive power in a complex world.

This predictive power goes a long way toward reconciling the increasingly baroque ontology of the various sciences with the idea of parsimony as a theoretical virtue. The ontology of modern cosmology may dwarf that of ancient Mesopotamia, but so too does its predictive power. The number, scope, generality, and precision of its predictions are incomparably greater. The complexity of modern cosmology, it seems, is more than made up for by its predictive power. However, why does a higher ratio of predictive power to descriptive complexity support any ontological conclusions about the postulates that facilitate those predictions?

To motivate this question, we will consider the eliminative materialist challenge to folk psychology, with special attention to its concerns about the propositional attitudes (e.g., beliefs and desires). Churchland (1981) defends this view but, notably, does not deny that folk psychology can leverage propositional attitude ascriptions to make highly successful predictions. He explains:

> The fact is that the average person is able to explain, and even predict, the behavior of other persons with a facility and success that is remarkable. Such explanations and predictions standardly make reference to the desires, beliefs, fears, intentions, perceptions, and so forth, to which the agents are presumed subject. (68)

Nevertheless, Churchland is skeptical about whether the subjects of these predictive efforts *really* have the propositional attitudes we ascribe to them. He outlines several reasons for this skepticism, but one is of special interest here:

> We must consider what sorts of theories are likely to be true of the etiology of our behavior, given what else we have learned about ourselves in recent history. That is, we must evaluate [folk psychology] with regard to its coherence and continuity with fertile and well-established theories . . . because active coherence with the rest of what we presume to know is perhaps the final measure of any hypothesis. (73)

He continues:

> We can tell a coherent story . . . of [human] constitution, development, and behavioral capacities which encompasses particle physics, atomic and molecular theory, organic chemistry, evolutionary theory, biology, physiology, and materialistic neuroscience . . . But [folk psychology] is no part of this growing synthesis. Its intentional categories stand magnificently alone, without visible prospect of reduction to that larger corpus. (76)

For Churchland, predictive power is all well and good, but ontological commitment to such things as beliefs and desires requires that they (and the larger theory of which they are a part) be reducible to more fundamental theories (esp. those of neuroscience). Without this, folk psychology is a useful fiction at best.[1]

The detailed requirements for reduction in Churchland's sense do not concern us here, and the intellectual landscape around these issues has changed greatly since 1981. What is relevant is that Churchland's worries arise from his view that folk psychology and neuroscience do not *hang together* in the right sort of way. In response, defenders of folk psychology must give an account of how they *do* hang together, one that allays concerns about the reality of the propositional attitudes. This account may or may not be reductive, per se, and it may require some revisions to folk psychology, but it must be clear about whether, and in what sense, the propositional attitudes deserve a place in our ontology.

Dennett develops just such an account across multiple works, and he uses the analogy of a *stance* to make clear how the propositional attitudes are deployed to facilitate prediction (Dennett 1971, 1989). In particular, when taking the *intentional stance*,

> one predicts behavior . . . by ascribing to the system *the possession of certain information* and by supposing it to be *directed by certain goals*, and then by working out the most reasonable or appropriate action on the basis of these ascriptions and suppositions. It is a small step to calling the information possessed the [system's] beliefs, its goals and subgoals its desires. (Dennett 1971, 90) (emphasis in original)

1. Churchland (1981) also argues that the predictive power of folk psychology is overstated, but we will focus on the more general concern that even highly predictive theories may misrepresent their target system.

For present purposes, we can think of intentional stance reasoning as the stance from which we apply folk psychology to predict behavior. Dennett's talk of "stances" emphasizes the possibility of adopting alternative stances toward a system. Indeed, Dennett identifies several alternative stances (e.g., the design and physical stance). Whether we take a particular stance, he contends, depends on both the predictive value of the stance and other "pragmatic" considerations (e.g., whether the predictions are useful or easy to infer)(1991, 90).

So far, Dennett's approach is consistent with viewing the propositional attitudes as useful fictions. Nevertheless, Dennett emphasizes that in many cases, the choice is *not* between a host of equally useful but ontologically incompatible stances. The intentional stance offers an unparalleled combination of simplicity and predictive power *for certain systems*—a feature that cannot be plausibly decoupled from facts about how the those systems are organized. Why else would the utility of the intentional stance vary from minimal for some systems (e.g., the planets in their orbits) to all but indispensable for others (e.g., humans or AI programs)? Given this, Dennett argues that any system whose behavior is "usefully and voluminously" predictable from this stance is an *intentional system*—a system with propositional attitudes (esp. beliefs and desires) which produces actions with an eye toward maximizing the expected satisfaction of its desires in light of its beliefs (2009, 339).

However, even if one accepts this argument, there is much work to be done in understanding these "differences in organization" and how they relate to the attitudes ascribed to agents in the course of intentional stance reasoning. To this end, Dennett proposes *real patterns* as a way of substantiating the connection between being usefully and voluminously predictable from the intentional stance and possessing an organization that substantiates propositional attitude ascriptions. We will address real patterns in two stages. First, we will step back and consider what kind of organization is required to be an intentional system. Second, we will consider why being usefully and voluminously predictable from the intentional stance is a necessary and sufficient condition for possessing such an organization.

1.2.2 The Organization of Intentional Systems

Before considering the organization of intentional systems, we should first consider a more mundane case in which we characterize the organization of a system. Suppose, for example, we are attempting to describe the functional organization of a particular archer's bow. Here, we may appeal to a range of considerations, from highly general principles of design or organization to the details of how particular design elements are physically implemented.

For example, we might first appeal to a highly general principle common to a range of functionally similar systems (e.g., store mechanical potential energy and release it to fire a projectile). We can then narrow this characterization by appeal to more specific design elements. For example, we might identify the use of a *compliant mechanism* (i.e., a mechanism that operates by elastic deformation rather than by joints connecting rigid parts) as the means of storing mechanical potential energy (Howell et al. 2013). Here we distinguish bows from *rigid* mechanisms for storing potential energy and launching projectiles (e.g., trebuchets).

We can further narrow our characterization by appealing to specific kinds of compliant mechanisms (e.g., a string and recurve bow). Here we distinguish the bow in question from differently curved bows (e.g., straight bows) as well as other projectile weapons that store mechanical energy via a compliant mechanism (e.g., onagers with their torsion mechanisms). Continuing, we can appeal to the finer details of implementation (e.g., the use of particular fibers or composite materials). These details distinguish the bow in question from similarly shaped bows. We can finalize our characterization of the bow by drawing together all these elements into an explanation of how the bow stores and releases mechanical potential energy to fire a projectile when operated by an archer.

In characterizing the organization of a particular intentional system, we can do much the same. As a broad principle of organization, we might say that an intentional system produces actions that tend to maximize the expected satisfaction of particular desires given particular beliefs. We might further specify that this is accomplished via "the manipulation of propositional attitudes" (Churchland 1981, 78). If we adopt this view, we might elaborate it with more specific design elements and (ultimately) implementation details. For example, we might specify that these propositional attitudes are represented by formal symbols and that the manipulation of propositional attitudes is accomplished via the manipulation of these symbols according to certain rules. These symbols, in turn, may be realized by specific vehicles (e.g., patterns of neural firing or bit strings in a digital computer). The manipulation of these symbols might be accomplished via machinery that operates on these vehicles (e.g., a neural network or a microprocessor) according to rules represented in specific ways (e.g., rules encoded in neural connections or programs represented as bit strings in a digital computer).

To see where problems begin to arise, note that at some point in our increasingly specific characterization of the bow, we stopped distinguishing bows from non-bows and started distinguishing this bow from other bows. A slingshot, for example, also uses a compliant mechanism to store mechanical potential energy and release it to fire a projectile. However, it cannot be characterized as having any particular bow geometry (e.g., straight vs. recurve). Thus, to identify something as a bow rather than something else requires knowing how a thing is organized with a nontrivial *but also nonmaximal* degree of specificity. For example, we may say that it requires a flexible bow and a string, but leave aside how the bow is curved, what the string is made of, and so on.

In determining whether something is an intentional system, we might very well disagree about where implementing an intentional system ends and implementing a *particular kind* of intentional system begins. Few would doubt that a system—whether a human brain or a general-purpose computer—that operates as described above is an intentional system. However, if human cognition is *not* well described as manipulating formal symbols, does that preclude our being intentional systems? That is, could activity worth calling the production of "actions that tend to maximize the expected satisfaction of particular desires given particular beliefs" be accomplished without formal symbol manipulation? Unfortunately, our ordinary definitions of terms like "belief" or "desire" are conspicuously silent on the allowable ways these propositional attitudes can be implemented. This is unsurprising since these terms came into usage before we had the faintest idea of how the brain might implement beliefs and desires.

Ultimately, it is a matter for philosophers and cognitive scientists to determine the appropriate level of specificity with which to define intentional systems. As it happens, eliminativists share key assumptions about this issue with what Dennett calls "industrial-strength" realists (1991, 30). Both assume a fairly strict (i.e., specific) set of conditions for being an intentional system, with the eliminativists being pessimists and the industrial-strength realists being optimists about whether human beings will satisfy these conditions. Dennett views both positions as misguided and proposes a more relaxed set of conditions for being an intentional system.

The strict view of intentional systems assumed by both eliminativists and industrial-strength realists has its roots in artificial intelligence, especially in work on the *physical symbol system hypothesis*. Newell (1980) describes the hypothesis as follows:

> This concept of a broad class of systems that is capable of having and manipulating symbols, yet is also realizable within our physical universe, has emerged from our growing experience and analysis of the computer and how to program it to perform intellectual and perceptual tasks. The notion of symbol that it defines is internal to this concept of a system. Thus, it is a hypothesis that these symbols are in fact the same symbols that we humans have and use everyday of our lives. Stated another

way, the hypothesis is that humans are instances of physical symbol systems, and, by virtue of this, mind enters into the physical universe. (136)

On this hypothesis, the idea of a physical symbol system captures the central organizing principle of the human brain as a cognitive system. Naturally, our particular physical symbol system would be one implemented by neurons rather than digital circuitry, but implemented it must be—it cannot be a mere interpretive gloss on the complexities of the human brain or human behavior. For this reason, we should expect to uncover a symbolic mental language upon which the human physical symbol system operates and in which our propositional attitudes are represented.

This hypothesis about human cognition can easily be transformed into a view about how *any* intentional system must be realized in order to justify realism about its supposed propositional attitudes. According to this view, an intentional system would be one in which beliefs and desires correspond more or less directly to elements of cognition that play functional roles analogous to their roles in folk psychology. For example, for beliefs to be real, there must be elements of cognition (e.g., symbols) that mediate the relationship between desire and action in the manner proposed by folk psychology (i.e., by constraining the set of possible actions to those that tend to maximize expected desire satisfaction).

On this view, to claim that humans are intentional systems with beliefs and desires is (again) no mere gloss on human behavior or neural processes. It is a claim about the cognitive architecture of human beings, though perhaps not one quite so specific as the physical symbol system hypothesis. Churchland captures this view clearly by noting that folk psychology must "describe faithfully the internal activities of a naturally distinct class of empirical systems" (1981, 78). From here, we can see where the eliminativist and industrial-strength realist part ways. The former (as we have seen) doubts that neurocience and psychology will ultimately rely on beliefs and desires to "describe faithfully" the functional organization of the brain. The industrial-strength realist, in contrast, holds that beliefs and desires will prove indispensable to such a description. For example, Fodor (1975) argues in favor of a *language of thought* on the grounds that computational models are the only plausible models of human cognition and that these computational models rely on a symbolic language (a kind of *mentalese*) for representing and manipulating the contents of thought.

Dennett's view substantially weakens the idea that to be an intentional system is to possess a specific kind of cognitive architecture. Instead, Dennett holds that the idea of an intentional system is a kind of regulative ideal rather than a principle of functional organization. In elaborating this idea, he invokes Anscombe's (1957) remarks on Aristotle's account of the practical syllogism:

> If Aristotle's account were supposed to describe actual mental processes, it would in general be quite absurd. The interest of the account is that it describes an order which is there whenever actions are done with intentions. (80)

Similarly, we might say that Bayes' theorem describes an order that is there whenever we make rational inferences from evidence, but that does not mean that we have made those rational inferences by utilizing Bayes' theorem explicitly. Indeed, many statistical methods that are not explicitly Bayesian admit Bayesian interpretations given reasonable assumptions about the world and our evidence (see, e.g., Wan 1990; Storey 2003; or McElreth 2016).

Thus, to be an intentional system on Dennett's account is to possess a functional organization that, to a close approximation, honors the norms of means-end reasoning. For example, consider the following maxim: "Take whichever action maximizes the expectation of immediate rewards plus the discounted expectation of future rewards." This good rule of decision-making is the foundational axiom of reinforcement learning, where this sum for an action in a state is called the *Q-value* of that action in that state (Sutton and Barto 1998). This maxim suggests *but does not imply* a strategy of decision-making on which we separately identify the expectation of immediate and future rewards, sum them for each possible action, and take whichever action has the highest sum. *Value iteration* agents do exactly this, combining prior knowledge of the reward function with a learned expectation of future rewards to calculate relevant Q-values. *Q-learning* agents learn Q-values directly and apply them with no explicit distinction between immediate and future rewards.[2] *Deep Q-learning* agents learn Q-values *implicitly* by learning weights for a neural network that maps a representation of the present state to a vector of Q-values for all possible actions.[3]

2. It is worth noting that Q-learning does employ a *learning rule* that (1) distinguishes between immediate and future rewards and (2) is derived from the decision rule discussed above. This kind of learning rule underlies a broader class of *temporal difference learning* algorithms of which Q-learning is a popular example. This gives us a more concrete sense of why the learned Q-values respect the decision rule even if Q-learning agents need not countenance the rule during actual decision-making. The learning process is derived from the decision rule and, in turn, determines the values the agent places on actions in each state.

3. To give an another example, suppose one's preferred spreadsheet software includes the ability to sort the values in a particular column from lowest to highest. There are multiple algorithms to accomplish this, but not all of them employ anything like a representation of "sorted from lowest to highest." For example, compare selection sort and bubble sort. Selection sort finds the lowest value and places it first, finds the second-lowest value and places it second, and so on. Bubble sort simply compares adjacent pairs of values in a list and swaps them if they are locally out of order. It

What we see in all these methods are different ways of honoring the central maxim of reinforcement learning, which vary in the degree to which they take the maxim literally as a design principle. It would be odd to insist that of all the agents discussed above, only value iteration agents should count as *real* reinforcement learners. In the same way, it seems strange to insist that a *real* intentional system will attempt to maximize the expectation of desire satisfaction given its beliefs via a particular kind of cognitive architecture. On Dennett's view, then, to be an intentional system is to exhibit a behavioral pattern honoring the norm of maximally satisfying a set of desires given a set of beliefs. These beliefs and desires, as elements of this pattern, are as real as the pattern itself.

There is much to recommend Dennett's middle way. First, it allows the graded instantiation of intentional systems. Different systems will exhibit this behavioral pattern to different degrees (e.g., by more or less reliably choosing desire-maximizing actions). Further, the advantages (in terms of simplicity or predictive power) of intentional stance reasoning will vary for different systems. For example, taking the intentional stance toward a thermostat is simple and predictive enough, but it offers no predictive advantages over a mechanical or algorithmic model. The same cannot be said of human beings.

Recognizing the graded instantiation of intentional systems gives Dennett's account striking explanatory advantages. For example, accounts of intentional systems with more stringent requirements make it harder to see how a system might have particular organization *to some degree*.[4] This undermines our ability to explain a number of well-supported scientific discoveries (e.g., the graceful degradation of neural systems or the gradual evolution of the human mind). It is easy to see how a system might more or less effectively translate

repeatedly passes over the list until no swaps are required. Selection sort explicitly involves global comparisons and the idea that the globally lowest value should be first. It does not explicitly attempt to find the globally second-lowest value, but this behavior is implicit in the fact that the first-lowest item is left out of subsequent comparisons. Bubble sort, on the other hand, does not involve any explicit global comparisons. Rather, the fact that it will sort the globally lowest value into the first position is entirely implicit in its systematic comparisons of adjacent values and its halting condition. As we can see, selection sort more perfectly encodes our understanding of a list of values sorted from smallest to largest. Nevertheless, it would be absurd to deny that bubble sort *really* sorts lists.

4. To briefly return to the reinforcement learning example, the Q-values of Q-learning agents converge with the true values over training, with a guarantee of convergence only as training time approaches infinity. If one observes agents learning over time, the degree to which they act rationally gradually increases.

beliefs and desires into actions. It is much harder to see how a system might more or less employ a formal language of thought.[5]

Second, isolating intentional systems as an object of scientific or philosophical study requires an account that is at least somewhat neutral with respect to implementation details. Here, a careful balance must be struck between the value of revising our concepts and the value of revising our ontology. We could adopt an account of intentional systems so *restrictive* about implementation that it resists revision in light of psychological or neurological discoveries. On the other hand, we could adopt an account of belief so *permissive* about implementation that no scientific discovery would warrant removing it from our ontology.

Both approaches present distinct risks and benefits. The risk of an excessively strict account is that we will discard a useful explanatory postulate (e.g., beliefs or desires) based on the discoveries about the minutiae of human cognitive architecture. The risk of an excessively permissive account is that no evidence could lead us to discard our commitment to beliefs, gutting their empirical significance and undermining realism about them. A more moderate position could (we should hope) avoid these extremes. It would allow us to remain open to discarding our beliefs in light of sufficient scientific evidence while remaining open to scientific discoveries about how beliefs are implemented. Thus, a middle way seems desirable—one that allows for graded implementation as well as revision or rejection in light of new evidence.

To borrow Dennett's turn of phase (1991), we should seek a "mild realism," intermediate between industrial-strength realism and eliminativism (30). Dennett proposes precisely this kind of account, requiring that intentional systems exhibit a real behavioral pattern while allowing flexibility about the cognitive architecture that gives rise to this pattern. Further, we can now appreciate in what sense Dennett affirms the reality of the propositional attitudes—they are element of a *real pattern* in the behavior of a system. However, we have not seen what makes patterns real or, more specifically, how the simplicity and predictive power of intentional stance reasoning is connected to the reality of relevant behavioral patterns. Resolving these issues is arguably where Dennett's account makes its most significant and wide-ranging contribution to the philosophy of cognitive science and the philosophy of science more broadly. We turn to this topic in the section that follows.

5. This problem has lead some defenders of a symbolic language of thought to embrace forms of saltationism. On these views, the basic elements of our capacity for symbolic reasoning evolved in "a single, rapid, emergent event" (Bolhuis et al. 2014).

1.3 Patterns, Information, and Compressibility

1.3.1 From Communication to Codes

Dennett (1991) connects the simplicity and predictive power of intentional stance reasoning to the reality of relevant behavioral patterns by appealing to a well-established result in information theory. This result is that the ability to compress data depends on the existence of patterns within that data. While this result is not controversial, its significance and justification are hard to appreciate without at least some relevant background. Dennett (1991) offers a clear and intuitive presentation of this result, which we hope to supplement here by considering the historical and conceptual foundations of information theory and of coding theory in particular.

How should we quantify the information content of a message sent via some channel of communication? Information theory tells us that the information content of a message depends principally on the probability of the event encoded by the message and the scheme used to encode it. It *does not* tell us that the information content of a message depends on facts about its sender or receiver (e.g., the intent or interpretation of the message). This is not an oversight but the result of an explicit effort by early information theorists to give an account of information and its communication that eliminates "the psychological factors involved" (Hartley 1928, 536).

Hartley (1928) observed that schemes for encoding messages often allowed for the production of gibberish. In particular, many encoding schemes allowed for the transmission of symbol sequences that failed to correspond to any meaning understood by relevant communicators. For example, international Morse code can be understood as consisting of five symbols: dot, dash, short gap, medium gap, and long gap. These symbols could be encoded, for example, by modulating the voltage along a wire (high or low) for different amounts of time, where each of the five symbols is encoded (respectively) by high for one time unit, high for three time units, low for one time unit, low for three time units, and low for seven time units. Morse code encodes characters (e.g., letters) by using sequences of dots and dashes separated by a short gap. Each character-encoding sequence is separated by a medium gap, and each word is separated by a long gap.

At both levels of interpretation (i.e., characters and words) it is possible to transmit sequences that do not correspond to anything the communicators would regard as meaningful. For Hartley, this was not a flaw in Morse code but an artifact of human psychology. For example, one character sequence could be meaningful for speakers of one language but not another. Similarly, the set of mutually intelligible sequences would vary for speakers of different languages or dialects. Of course, no consistent measure of information-carrying capacity could depend on who was communicating what with whom or which language(s) they chose for the purpose. Fortunately, Hartley made a critical observation. Hartley recognized that an uncomprehending telegraph operator could recognize the basic symbols of an encoding scheme without any ability to interpret them.

Taking the perspective of this uncomprehending operator, Hartley proposed a measure of information-carrying capacity that depends only on the number of symbols sequences (i.e., messages) possible under the relevant encoding scheme. For encoding schemes that employ sequences of a fixed length, n, and a set of symbols of size s, the number of possible sequences is given by s^n, and Hartley's measure of information for that channel would be $\log s^n$. Intuitively, the reason for considering the number of possible messages is that it tells us how many distinct events could be communicated by a single message. For example, if the possible sequences are $\{0, 1\}$, we could communicate the outcome of a coin flip in a single message. If the possible sequences are $\{0, 1, 2, 3, 4, 5\}$, we could communicate the outcome of a die roll.

The reason for taking the logarithm of possible sequences is to ensure that the information content of transmission grows linearly with both the number of sequences sent, m, and the sequence length, n. This means, for example, that *either* (1) using twice as much paper to write down m sequences of length $2n$ or (2) using twice as much paper to write down $2m$ sequences of length n will double the information contained on the paper. Similarly, using either (1) twice as much time to send m sequences of length $2n$ or (2) using twice as much time to send $2m$ sequences of length n will double the information sent over the time interval. This links information content directly to the capacity of physical media and the time available to transmit data.

Hartley's (1928) foundational work in information theory makes a critical assumption in its effort to de-psychologize information. It assumes that the choice of what to transmit is "perfectly arbitrary" (538). For example, he assumes (for the purpose of his account) that the string "sdgljadf" is just as likely to be transmitted as the string "shortened." This assumption is stronger than it needs to be to accomplish the purpose of de-psychologizing information theory. Consider a system for encoding letters in binary and transmitting them via some channel of communication. For any real-world communication channel, we can *without comprehension of their significance* learn the statistics of letter transmission. We could, for example, note that the "e" symbol is transmitted more than the "q" symbol. These statistics depend on those sending the messages via the channel but *not* in a way inaccessible to the uncomprehending operator.

Along these lines, Shannon (1948) assumed a space of messages (e.g., characters or words) that we might want to encode *and* a probability distribution over that space. The relevance of this observation is not immediately obvious. If every message is encoded with a fixed-length sequence of symbols, then their probabilities of transmission will not affect, say, how many messages can be transmitted in a given time or recorded in some physical medium. In contrast, we might wonder whether using shorter sequences for more frequent messages and longer sequences for less frequent messages might allow us to say more in less time or to write more in less space. As it happens, this strategy does work, and we can quantify the number of symbols we should use under an optimal encoding scheme to encode any particular message based on its probability.[6]

Shannon did not provide an algorithm for creating optimal encoding schemes.[7] Rather, he gave a formal theory of information that placed an inclusive lower bound on the average code length for messages drawn from a given distribution under an optimal encoding. For binary codes, this lower bound is as follows:

$$\sum_{0}^{N} \log_2 \left(\frac{1}{P(m_n)} \right) P(m_n). \tag{1.1}$$

Here $P(m_n)$ is the probability that the *n*th message in the set of possible messages is sent at any given time. This means that the contribution of any particular message, m_n, to the average will be $\log_2(\frac{1}{P(m_n)})$ weighted by its probability. This also implies that the code length of a message under ideal circumstances will be equal to $\log_2(\frac{1}{P(m_n)})$.

The proof of this result is beyond the scope of this introduction, but we can make this result more intuitive if we consider the following scenario. We and a friend are going to play a guessing game. Our friend will draw a marble from an urn, and we will try to infer the color by asking a series of yes or no questions. This urn contains marbles of six distinct colors. There may be more or less of any particular color, but the fraction of the marbles corresponding to each color is fixed and will not change from game to game. What is the best strategy for playing this game?

In the related game *Guess Who?* it is widely known that the best strategy is to ask yes or no questions that rule out half of the remaining options. Fortunately, characters in the game have many distinguishing features, and it is often possible to construct a query that will follow this optimal strategy (i.e., by asking

6. See chapter 5 for further details on this point.

7. For such an algorithm, see Huffman (1952).

about a trait that exactly half the remaining characters possess). If we cannot follow the strategy exactly, we might choose whichever question would *most equally* divide the remaining characters. This is not quite the optimal strategy, but it is a good one.

The marble game differs from this game in that we are trying to guess the color of the marble chosen rather than the particular marble chosen. This has interesting consequences for the set of questions we should ask. We will continue to select questions that most equally divide the remaining marbles into equal subsets. For example, suppose red and blue each make up one-quarter the marbles while the remaining colors each make up one-eighth of the marbles. In this case, our first question might be "Is the chosen marble red or blue?" Whether the answer is "yes" or "no," we will eliminate half the marbles from consideration. While either answer eliminates the same fraction of marbles, the answers have different implications for how many questions remain. If the marble chosen is red or blue, then we need ask only one more question ("Is it red?" or "Is it blue?"). If the marble chosen is neither red nor blue, then two questions will be required, the first to eliminate two of the remaining four colors and the second to identify the exact color chosen.

Given this, it is easy to see what condition must be satisfied to play our chosen strategy perfectly. The fraction corresponding to each color must be expressible in the form: $\frac{1}{2^n}$, where n is an integer. Since the fractions must sum to one, we will always be able to divide the remaining marbles into two equal subsets. Further, in this ideal case, the number of questions we must ask will be: $\log_2(\frac{1}{p})$, where p is the fraction of marbles of the chosen color in the urn.

This is exactly the result shown above for binary codes, but why? The reason is that, in both cases, the thing to be communicated (i.e., the marble or message) is indicated by a sequence of binary signals—*zero* or *one* in the first case and *yes* or *no* in the second. The only thing needed to make the cases completely identical is to determine ahead of time *with our partner* which question we will ask first and which questions we will ask next in response to prior answers. Once we have done this, our partner can simply recite a sequence of "yes" or "no" answers to indicate the color drawn. No questions actually need to be asked since our question-asking plan is known to both of us—not unlike a previously agreed-upon code. Further, if the marble fractions are ideal (as described above), then the length of these sequences will be (1) as short as possible and (2) inversely proportional to the color frequency of the chosen marble.

The fundamental idea in both cases is that we can minimize the expected number of bits required to communicate a series of events by assigning shorter codes to more probable events and longer codes to less probable events. As a result, where our distribution is more regular (i.e., transmits one or few messages nearly all of the time), we will send shorter codes on average since most of the messages sent are of relatively high probability. Where our distribution is less regular (i.e., transmits many different messages with near equiprobability), we will send longer codes on average since all messages are of relatively low probability. This helps us to see how data that exhibits a distinct pattern (i.e., greater regularity) could be encoded using fewer symbols (i.e., *compressed*) whereas data that exhibits no pattern or a weak pattern (i.e., less regularity) would require more symbols to encode. The connection is more than merely intuitive. Equation 1 is simply the *entropy* (a measure of disorder) of the message source in units of bits if we treat the messages as values sampled from a random variable.

1.3.2 From Algorithms to Models

The connection can be made even more vivid by considering compression algorithms.[8] Compression algorithms take data, represent it in a more economical form, and recover the original data when needed. In some cases, compression is *lossless*—the original data can be recovered *exactly* from the compressed representation. In other cases, compression is *lossy*—the original data can be recovered *approximately* from the compressed representation. As before, when (and only when) data exhibits a pattern is it possible to represent that data more efficiently and, later, to recover our original data from that representation—whether perfectly or to a close approximation.

A compression algorithm, for example, might take a text document and use character frequencies within the document to assign each character a binary code—with more frequent characters getting shorter binary codes and less frequent characters getting longer binary codes. However, many other approaches are possible. For example, the algorithm might detect strings of repeated characters and replace them with something equivalent to *repeat "234" one hundred times*. Subtler methods (such as JPEG compression) might sacrifice accuracy with respect to fine details to better capture coarse ones. Still other methods might take advantage of correlations between elements in data sequences. For example, interframe compression algorithms compress video by encoding only how a frame differs from the preceding frame rather than starting from scratch each frame. This is why portions of streaming videos with

8. Here we transition from classical information theory to algorithmic information theory. For more information about the differences between these approaches to information, see Grünwald and Vitányi (2008).

few moving objects exhibit few compression artifacts, but portions with many moving objects (e.g., snow or confetti) exhibit many compression artifacts.

In each of these cases, the success of the algorithm is dependent not merely on patterns but on distinctive assumptions about the patterns present in the data. If all characters are about equally likely, then assigning shorter characters to more frequent characters yields little benefit in terms of efficiency. If characters don't repeat much more often than you'd expect by chance, then encoding runs of characters won't save many characters. If an image is just random noise and lacks any coarser-grained patterns, JPEG compression will struggle to recover the original image with any accuracy. If a video changes wildly from frame to frame, then interframe compression will have to effectively start from scratch with each frame, reducing the efficiency of only encoding differences between frames. Again, the point here is that compression depends not only on patterns in general but on patterns that comport with the basic assumptions of the compression algorithm.

For Dennett, this is the key fact connecting the success of intentional stance prediction to the reality of relevant behavioral patterns. One might attempt to create a highly detailed model of the human brain at a neurophysiological level and use this model to make predictions about behavior. It may even make highly accurate predictions, but this will come at the cost of extreme complexity. We can still do surprisingly well, however, by taking the intentional stance toward the system. Here we sacrifice *some* predictive accuracy, but with the benefit of a much simpler description of the system as an agent with a collection of beliefs and desires.

There is clearly an analogy here between the predictive economy of the intentional stance and the descriptive economy of a good compression algorithm. There is some challenge in making this analogy precise (Millhouse 2021), and the best way to understand it is a controversial matter—whether one considers the question from a philosophical or interpretive perspective. We will attempt to make this analogy more precise without unduly taking sides in these debates by comparing an example from Dennett to an example of our own.

Suppose we are predicting the behavior of a computer running a chess-playing program. With enough patience, we might actually be able to give a fairly complete low-level description of the computer, whether in physical or software terms, but both of these require a detailed understanding of the state of the system. We might instead take the intentional stance toward the system, attributing to it a belief about the state of the board and the goal of winning the game. This will allow us to predict the next state of the board from the current state of the board with remarkable accuracy whenever it is the program's turn to play. This accuracy is remarkable not so much because it precisely predicts our digital opponent's play, but because (1) our assumption rules out so many possible moves on strategic grounds and (2) our assumption says so little about the state of the chess-playing program.

Now consider a hypothetical interframe compression algorithm for video. This hypothetical algorithm doesn't merely encode the differences between one frame and the next—it uses an AI model for next-frame prediction and only encodes the error between the prediction and the next frame. When it is time to decode the video, the algorithm (1) uses the model to *predict* the next frame, (2) corrects the prediction (based on the recorded errors), and (3) displays the frame. We will assume that the algorithm and model are encoded along with the compressed file (i.e., record of errors), so any computer can decode the video without additional software.

It is easy to see that the simpler or more accurate the model is, the smaller the resulting file will be since either the included model or the included file (i.e., recorded errors) will be reduced in size. Further, it is clear that the model will only be able to efficiently predict next frames if there is a pattern relating each frame to the next. If there is no pattern, the model will either need to memorize next frames (increasing its size) or sacrifice accuracy (increasing the amount of error information encoded in the file). Finally, whatever pattern is present, it must be consistent with the presuppositions of our model if our model is to be at all accurate.

Our prediction of the next state of the chessboard seems highly analogous to this situation. The better our simple intentional systems model of the program predicts next board positions by predicting the behavior of the program, the smaller our predictive error will be. Of course, the success of such a simple model is only possible if the behavior of the system *really does* exhibit the kind of pattern presupposed by our model. Indeed, this pattern is just the kind of pattern discussed earlier—one in which the behavior of the system (to a close approximation) honors the central assumption of the intentional stance—that the system attempts to maximize the satisfaction of its desires in light of its beliefs.

1.4 Beyond Beliefs

The subsequent development of Dennett's idea of *real patterns* has taken two primary courses. First, real patterns have been employed (often in a precisified form) to address questions in the general philosophy of science, especially to questions about realism, emergence, and reduction. As yet, there are comparatively few applications in the philosophy of particular sciences outside of the cognitive sciences—though such applications certainly exist (see, e.g., Burnston 2017) and several appear in this volume (see, e.g., chapter 9). Second, real patterns has been a subject of philosophical study in its own right, with various authors attempting to analyze and amend the view—often with an eye toward clarifying or strengthening its realist credentials. We will briefly address work in both of these areas in the subsections that follow, but it is worth pausing to say something more general about the real patterns literature.

Work on real patterns has been (and continues to be) highly pluralistic, marked by many variations and few rivalries. Perhaps this is due to a sense of common cause arising from a common inspiration. There is certainly a factor, but we would also point to a feature of Dennett's approach itself. In "Real Patterns," Dennett (1991) closes with this preemptive riposte to his future critics:

> Is the view I am defending here a sort of instrumentalism or a sort of realism? I think that the view itself is clearer than either of the labels, so I shall leave that question to anyone who stills find illumination in them. (51)

Arguably, Dennett overestimated the clarity of his view and underestimated the number of readers who would still find the question of realism and instrumentalism illuminating. Nevertheless, this passage captures something essential about those who have been inspired by Dennett's insights.

This essential element is a focus on articulating a view that is clearer than any controversial philosophical category. This focus means that even if one worries that a variant of real patterns is insufficiently (or excessively) realist, the worry rarely extends to the coherence or value of the underlying view. Most variants of real patterns identify a clear and compelling framework for thinking about patterns in the world—a framework that *at the very least* is worthy of careful explication and analysis. For this reason, proponents of real patterns can appreciate the distinctive contributions of rival accounts while continuing to pursue any philosophical questions they find illuminating.

1.4.1 The Philosophy of Science

"Real Patterns" has amassed almost two thousand citations since its publication in 1991. As such, we cannot hope that this survey will be fully comprehensive of the subsequent literature. We will, however, attempt to make it illustrative of that literature, highlighting major themes and representative contributions. To begin, we will discuss the efforts by Ross, Ladyman, and others to adapt Dennett's real patterns into a general account of nonreductive physicalism and emergence suitable for the scientific metaphysician. These authors defend a mild, pattern-based realism closely aligned with Dennett's own.

We will first contexualize this work by briefly discussing two early critics of "Real Patterns"—the eminent philosophers John Haugeland and Richard Rorty (Dahlbom 1993). We will return to one of those critics (Haugeland) in greater detail in the section that follows. Both Haugeland and Rorty welcome the general thrust of Dennett's essay but criticize it in complementary ways. Haugeland argues that the essential point of real patterns is that we cannot understand patternhood as prior to or separable from recognizability. This necessarily implies a potential recognizer. Rorty amplifies this verificationist point—noting that Dennett defends what he calls "urbane verificationism"—in accusing Dennett of pulling his punches on two general philosophical points (198).

First, Rorty argues against Dennett's appeal to Reichenbach's distinction between *illata* and *abstracta*. Patterns that are useful for coping with the world, including developing scientific models of its aspects, are not generally abstractions from more basal concrete particulars because we often have no access to the particulars as the particulars they are except by way of their participation in patterns. For example, some physical elements are modeled as "particles" because of the role they play in successful models in physics; scientists do not first identify them atheoretically as particles and on that basis go on to construct "particle physics."

Second, Rorty argues against Dennett's framing of his realism as a "mild" goldilocks point between "industrial-strength" realism that seeks an ontology wholly insensitive to the pragmatic interests of the ontology's users and a mild (as attributed by Dennett to Rorty) irrealism according to which nothing should be said to exist except in the eyes of some actual community of beholders (Dennett 1991, 30). Reality, Rorty urges, does not come in degrees: Scientists treat many sorts of things—bosons, fields, chemical bonds, tectonic plates, short-term interest rates, and gendered expectations—as real, and various other things sometimes taken to be real—witches, the luminiferous ether, the Cartesian theater of consciousness, a Jewish conspiracy to rule the world—as not real at all. There is no third category of "semi-reality." This is a point about how inquirers use the idea of reality, not a point of first-order metaphysics. "Real" patterns, then, should not be regarded, according to Rorty, as fuzzy composites assembled out of more industrially real components.

These closely related themes lay idle in the philosophical literature for a few years, until, in another collection of essays on Dennett's work (Ross et al. 2000), Don Ross explicitly drew their implications together. Rorty's critique, Ross argues, implies that in a naturalistic ontology Quine's famous slogan "To be is to be the value of a bound variable" should be replaced by "To be is to be a real pattern." This has the portentous implication, as Haugeland had stressed in his 1993 essay and we stress above, that science is not generally in the business of clearing away folk ontological overgrowth in favor of an austere "desert landscape" of most general forms of being, but rather of continuously widening our sense of the world's complexity; we get a "rainforest realism."

However, Ross also goes beyond Rorty's point in recognizing that it is useless to complicate accounts of the world merely by recombining their elements in proliferating labels. We should adopt an ontologically serious attitude only to the patterns that scientific development finds to be necessary for empirically adequate modeling. He derives a limiting principle from Dennett's appeal to information compressibility. He thus motivates a two-part statement of natural ontological theory that he calls "Rainforest Realism" (RR):

> To be is to be a real pattern; and a pattern is real iff
>
> (1) it is projectible under at least one physically possible perspective; and
>
> (2) it encodes information about at least one structure of events or entities S where that encoding is more efficient, in information-theoretic terms, than the bit-map encoding of S, and where for at least one of the physically possible perspectives under which the pattern is projectible, there exists an aspect of S that cannot be tracked unless the encoding is recovered from the perspective in question. (Ross 2000, 161)

The first element of this theory, projectibility, captures the idea that real patterns must be genuine data-generating processes—including, but not necessarily, causal processes—and not just accidents of correlation. The most pervasive such accidents, with which econometricians constantly wrestle, are those that are maintained simply by the participation of many processes in the same general arrow of time. The reference to physically possible perspectives incorporates Haugeland's point that patterns must be recognizable, entailing specification of a class of recognizers, while also avoiding limiting ontological commitment to those patterns that have in fact recognized, or to patterns that humans, with limited cognitive capacities, could recognize (see below). Crucially, physically possible discernment is a modal criterion, which is why, notwithstanding Ross's endorsement of Dennett's "urbane verificationism" and Rorty's pragmatism, he incorporates "realism" in the name of the theory. Finally, clause (2) expresses the restriction that real patterns must be necessary to support some specifiable scientific inferences, and not just trivial concatenations of other real patterns where the epistemic gain from tracking the whole adds no gain to the sum of the parts. Note that Ross's formulation allows for lossy compression, since gains could involve epistemological trade-offs on margins.

Clause (2) is obviously under specified from an information-theoretic and computational perspective, leaving "efficiency" as a black box. Subsequently, Ladyman and Ross (2007, chap. 4) aimed to inject at least some content into this space. Following a review of varying uses of information-theoretic efficiency concepts in physics, evolutionary theory, and computer science, Ladyman and Ross refined Ross's original formulation of rainforest realism as follows:

> To be is to be a real pattern; and a pattern $x \rightarrow y$ is real iff
>
> (1) it is projectible; and
>
> (2) it has a model that carries information about at least one pattern P in an encoding that has logical depth less than the bit-map encoding of P, and where P is not projectible by a physically possible device computing information about another real pattern of lower logical depth than $x \rightarrow y$. (Ladyman and Ross 2007, 233)

The appeal to logical depth here involves a basic trade-off typical of theory development. On the one hand, it adds specification to the original formulation of rainforest realism, making it clearer. On the other hand, this makes the theory less likely to be generally empirically adequate, since alternative specifications of informational efficiency might select a broader, or narrower, or cross-cutting, set of real patterns. These alternative specifications could all, in principle, be presented as variants on the original statement of rainforest realism.

The Ladyman and Ross specification in terms of logical depth is standardly taken as a baseline version of the theory in the subsequent literature on Dennett's compressibility criterion for pattern reality, which we go on to review. But first we note that this issue is strictly separable from the wider metaphysical ideas that mainly motivated the more general statement of rainforest realism. This theme has generally been underexplored in the critical literature on Ladyman and Ross's very widely cited work. In particular, critics preoccupied with the computational and information-theoretic issues have often incorporated assumptions—for example, that data registration must precede pattern recognition, or that compression should not be lossy—that are incompatible with the empiricist and verificationist commitments underlying rainforest realism. For example, Wallace (2012) explicitly uses a specific information-theoretic version to support a version of realism about quantum entanglement and interference that, he argues, forces commitment to the

Everettian multiverse. Ladyman and Ross (2013) contest this inference by reiterating their verificationist and (Peircean) pragmatism about the ontological status of real patterns.

1.4.2 Real Patterns Revised

Dennett was wisely silent on several of the technical details of his proposal. It would arguably have been a mistake for him to wade into the minutiae of different information-theoretic measures of simplicity or compressibility (e.g., logical depth vs. Kolmogorov complexity). By giving a more schematic account, Dennett was able to capture an intuitively appealing idea (i.e., that many of the things studied by science are well understood as patterns) *and* avoid tying his view too closely to any specific formulation. As we have just seen, this allowed ample room for subsequent authors to fill out these details while hewing closely to the spirit of Dennett's original view. At the same time, the schematic nature of Dennett's view has left others uncertain of, or unsatisfied with, its realist commitments. In what follows, we briefly survey attempts to either clarify or modify those commitments.

One important ambiguity within real patterns was noted not too long after Dennett's original work. In the essay cited earlier, Haugeland (1993) points out that Dennett seems to speak of "patterns" in two senses, as both things or events that exhibit a particular regularity and as the regularity itself. To borrow one of Dennett's prime examples, there are the spaceships and other structures in the Game of Life and there are the behaviors they exhibit (e.g., the behavior of an eater interacting with a glider). Both seem necessary to understanding how we might make successful high-level predictions about the Game of Life.

This kind of ambiguity seems ontologically innocent. Patterns, like causes, may be spoken of in many ways. Haugeland (1993), however, suspects that this ambiguity prevents us from fully appreciating Dennett's distinctive "ontological insight" (476). Haugeland distinguishes elements of patterns that are candidates for pattern recognition *in their own right* (e.g., the configuration of cells in one stage of a glider) from the case of elements whose recognizability is inextricably linked to their *participation in a pattern* (e.g., a glider that persists indefinitely through changing configurations).

Haugeland is concerned that if we fail to recognize this distinction and think of pattern elements *only* as candidates for pattern recognition in their own right, we will be left with an account of what it means for a real thing to "count as" an element of a pattern rather than an account of what makes a thing real *qua* pattern. He notes:

> The greater usefulness, in scientific calculation, of centers of gravity does not confer on them any further ontological status; they are no more "real" than lost-sock centers—just more worth keeping track of. This is not at all analogous to the claim about beliefs . . . Dennett never suggests that beliefs are simply a subset of the set of all brain structures (or computational states, or anything else independently specifiable) which, by virtue of participating in some pattern, also count as intentional. (1993, 55)

As these remarks illustrate, Haugeland is not ontologically impressed by the identification of things worth tracking. Rather, he is interested in extracting from real patterns a distinctive ontological account of things (like beliefs) whose ontological status *does* depend on their participation in a pattern, and he does this by identifying a distinctive mode of recognition that depends (at least in part) upon the recognition of a larger pattern in which the recognized thing participates.

Others have been more impressed by (and more vexed by) the identification of things worth tracking. This builds on another element of Dennett's original piece—his distinction between "real" and "bogus" patterns (29). Here, the distinction is not so much between things that are not real and things that are real, but things that are real and things that are *nontrivially* real. There are two broad approaches to this distinction—one that grounds nontrivial reality in a thing's participation in (or illumination of) a real pattern and one that grounds nontrivial reality in the individual recognizability of the thing in question.

The former view aligns with previous thinking about what it is that makes a kind *natural*. Take, for example, Fodor's (1974) view that "the natural kind predicates of a science are the ones whose terms are the bound variables in its proper laws" (102). Here, the naturalness of a predicate doesn't depend on the predicate itself (e.g., on which microstates are in its extension) but on whether that predicate figures appropriately in the formalization of a law-like regularity. However, insisting on understanding natural kinds in terms of laws ignores the many other ways that science represents regularities in the world (Godfrey-Smith 2006). Fortunately, real patterns provides a more general way to identify genuine regularities in the world, regularities whose elements are non-bogus by association.

One example of this approach to the elements of real patterns is clearly seen in Andersen's (2017) treatment of causal relations under various coarse-grainings of the phase spaces of physical systems. There are many ways to coarse-grain these spaces and only some of those ways support predictive modeling (e.g., causal modeling). She states:

> The level at which one describes a system corresponds quite closely to the grain of the partition . . . By varying the grain of partitions over two candidate causal relata,

> we vary the amount of mutual information between them. We can use this fine-tuning to discover proportional and maximally specific causal relationships by looking for partitions that maximize the mutual information between causal relata. (613–614)

Here we can identify non-bogus causal relata via coarse-graining by asking whether the resulting coarse-grained relata carry information about one another.

This frames the issue quite technically and in terms more familiar to physicists, but the underlying issue here appears in many domains. For example, should we characterize a study subject's behavior as *smoking tobacco* or *smoking Marlboros* when developing a causal model of cancer risk factors? Barring extraordinary information about the contents of Marlboro cigarettes, characterizing the behavior as *smoking Marlboros* seems likely to obscure an important causal regularity. This example is explored in greater detail by Kinney (Kinney 2019; Kinney and Lambrozo 2022), who also proposes an information-theoretic criterion for selecting the right level of grain in causal modeling.

In a more pragmatic mode, K'astner and Haueis (2021) articulate an account on which the nontriviality of an element and the role of that element in a pattern are connected by scientific practice. They remark:

> Scientists adopt epistemic perspectives to constrain their search for orderly arrangements that are stably recognizable. The subsequent discovery of the elements of such patterns involves recognizing that the pattern is there (outer recognition) as well as recognizing how a given element fits that pattern (inner recognition). (1646)

Here, scientists are seen as building upon the perspective, or *stance*, that enables them to recognize regularity in need of explanation. Scientists then engage in a process that involves the development of a refined perspective that both (1) permits the recognition of constituent elements (e.g., particular parts whose interactions give rise to the regularity) and (2) better isolates and confirms the larger patterns as an appropriate target of scientific inquiry. Ultimately, the nontrivial reality of these elements is grounded in their value in mechanistic explanation discovered in the course of this mode of scientific practice. As such, for K'astner and Haueis, the reality of these elements is *practice-relative* in a way not countenanced by Andersen and others.[9]

As mentioned before, another approach to the elements of patterns focuses on elements as candidates for pattern recognition in their own right. This approach also aligns with earlier work. Sometimes the focus is practical or

9. Andersen (2017) and Kinney (2019) are both open to *pragmatic* considerations playing some role in identifying non-bogus pattern elements, but these are primarily linked to the interests and goals of scientists rather than the practices of their scientific community.

epistemic. For example, Wimsatt (1981) articulates a concept of *robustness* that emphasizes independent identification and verification. Other times it is metaphysical. For example, varieties of reductionism are often characterized by the kind of mapping required between lower and higher levels of description (Fodor 1974). One might view any such mapping as an account of how to recognize things at the higher level given a complete description of things at the lower level. For example, if a system is in higher-level state *A* if and only if it is in a lower-level state of type *B*, then we can recognize *A* by checking whether the lower-level state is of type *B*. Of course, if one insists that a sufficiently demanding sort of mapping is needed for higher-level things to be nontrivially real (see, e.g., Churchland 1981), one might be less-than-fully receptive to the relative ontological permissiveness of Dennett's mild realism. For those interested in constraining allowable mappings *but sympathetic to Dennett's mild realism*, the challenge is to identify an appropriately mild constraint on mappings.

This challenge is closely related to a long-standing problem in the philosophy of computer science concerning computation in physical systems. Here, perhaps more than anywhere, there is a serious appreciation for the challenges of identifying non-bogus pattern elements. Triviality arguments against physical computation (e.g., Putnam 1988) attempt to show that ascriptions of computation to physical systems are trivial. One type of triviality argument attempts to identify simple systems that support at least one (but typically vastly many) ascriptions of complex computational activity. Millhouse (2019) calls these arguments "rock" arguments in reference to Chalmers's (1996) paper "Does a Rock Implement Every Finite-State Automaton?"

At their core, these triviality arguments turn on the flexibility of computational interpretations. These interpretations also take the form of *mappings* that map relevant states of a physical system (e.g., a rock) to computational states of an abstractly specified machine (e.g., a particular finite-state automaton). Some constraints on these mappings concern the behavior of the system *under the mapping*—that is, under a given interpretation. For example, it is easy to map an arbitrary sequence of physical states to an arbitrary sequence of computational states, but it is more difficult to construct a mapping under which the system satisfies relevant counterfactuals at the computational level (e.g., that were the input to the finite-state machine different, it would have entered a different state). Unfortunately, it is still too easy to satisfy this counterfactual constraint, and simple changes (e.g., introducing the ability to write on the rock) substantially undermine its force.

This led Millhouse (2019) to impose an additional constraint on such mappings and later (2022) to generalize this constraint to *any* mapping from states of a physical system to states of a higher-level model. This *simplicity constraint* complements the counterfactual constraint and quantifies the *bogusness* of the elements of a pattern (e.g., machine states or variables modeled) as the complexity of this mapping, where the complexity of the mapping is the length of the shortest program capable of computing the mapping on a reference universal Turing machine (i.e., its Kolmogorov complexity). As it turns out, thinking about such a program helps us to understand what work must be done to implement a mapping and, in turn, whether the supposed computation is actually carried out by the system or whether it is merely an artifact of our interpretive efforts. In the case of physical computation, it turns out that a good way to shorten the program for a spurious mapping (but not a legitimate mapping) involves replicating the supposed computation itself.[10] This is a sign that the real work of computation is being done by our mapping rather than the target system.

It would require a separate treatment to fully situate this kind of view on Dennett's spectrum of realism (i.e., from eliminativism, to mild realism, to industrial-strength realism). However, note that this approach to assessing pattern elements weakens the constraints on mappings imposed by strong forms of reductionism. For example, it entirely skirts concerns about multiple realizability across systems by permitting a bespoke mapping for every system that implements a particular abstract machine. We need not say *in general* what it means for a physical system to implement a particular machine (or any other higher-level model). We need only say what it means for *this or that* physical system to implement a particular machine. High-level states can also be disjunctive, so long as they can be recognized at the lower level by a relatively simple program. This kind of view (in essence) places highly general constraints on the stances we might take toward a system. This allows maximal flexibility in how pattern elements are recognized from a stance while also reassuring us that our stance is not imposing patterns on a system where none exist.

10. For example, the ability to write on the rock (perhaps by carving letters into the surface) means that the rock can enter as many distinct physical states as there are possible input strings that fit on the rock. This allows a mapping to be constructed that maps each of these many states to the correct machine state without ambiguity and in a way that supports all relevant counterfactuals (up to some limit on the rock's "memory"). An efficient program for computing this mapping might simply read the writing off the rock and pass this input to a subroutine in the program that implements the finite state machine attributed to the rock under the mapping. Indeed, it's hard to see how one could improve on this approach while omitting the subroutine. An input $\rightarrow$ state look-up table, for example, would be vastly more complicated for a finite state machine of any complexity.

Another issue for real patterns is the possibility of *redundant* patterns. Most advocates of real patterns are happy to admit that systems exhibit multiple, equally real patterns, but certain patterns seem redundant in a way that strikes some as degenerate. For example, suppose a model predicts a set of dependent variables perfectly given a set of independent variables. It will often be possible to complicate such a model without changing its predictive accuracy (e.g., by adding additional terms with zeros as coefficients). The resulting model is different from and inferior to the first, but some might want to go further and say categorically that the first captures a real pattern and the second does not.

For example, Suñé and Martínez (2021), building on a proposal in Ladyman and Ross (2007), require that real patterns be optimally simple and informative about their target data. There may be more than one optimally simple and informative real pattern, but redundant patterns of the kind described above will be excluded. In contrast, Petersen (2019, 2023, this volume) allows *any* compressing regularity to count as a real pattern in data. On this account, redundant patterns will simply be less significant in proportion to their added complexity. Ultimately, these accounts differ in whether to tie reality to nonredundancy—that is, whether only the most efficient patterns should count as real. To many Dennettians, of course, this question may prove to be a pragmatic one.

1.5 Our Contribution

This volume includes explorations of both of these approaches to real patterns. Some chapters are mainly concerned with how to extract patterns that can facilitate scientific inferences under specific domain restrictions. Other chapters are more concerned with what discovered patterns tell us about the structure of the world, considering factors that make some patterns more ontologically significant—more "real"—than others. It might be supposed that any discussion of the second issue is relevant only to philosophers interested in (naturalized) metaphysics. However, some chapters belie this assumption, situating problems of ontology specification in specific scientific contexts, including biology, economics, linguistics, and AI.

We open the collection with a reprint of Dennett's original paper, the wellspring of everything else in the book. We envy those readers who encounter it here for the first time, for they will get maximum enjoyment from what we consider one of the classic essays of twentieth-century philosophy. Readers already familiar with "RP" might benefit from the convenience of a review before digging into the subsequent elaborations on the challenges and opportunities stirred up by Dennett's insights.

The organizational logic of the volume is to begin with the most abstract framings and then move to more specific scientific settings. Ladyman addresses the major metaphysical themes, then Petersen refines the computational ones; both general faces of Dennett's idea are thus brought up to date. Millhouse's chapter consolidates the two themes into a proposed update of Dennett's general aim.

The three chapters that follow form a topical set. One common reading of "Real Patterns" is as an account of emergence in scientific ontology. Carroll and Parola, in their chapter, specify forms of emergence as might be reflected in physical theories. McAllister offers a skeptical perspective on the idea that any patterns are more ontologically significant than any others, even if some are obviously more relevant to scientifically important inferences. Wallace also addresses these general themes, offering a striking diagnosis of what, in general, goes wrong when philosophers try to extract ontological distinctions from scientific models that seek greater precision than the mathematics imply. This leads not to McAllister's skepticism about the potential of some patterns to matter more than others with respect to describing the structure of the world, but rather to modesty—arguably, closely reflecting and refining Dennett's attitude—about the extent to which a scientific ontology based on patterns can be used to vindicate any traditional metaphysical doctrine expressed using the methods of logical analysis.

These chapters on general themes are followed by applications. Hermida puts the idea of real patterns to work in fashioning a fresh take on long-standing debates in biology over the grounds for taxonomic and phylogenetic classifications. In a similar spirit, Alekseev, Harrison, Lau, and Ross show how another core idea of Dennett's, the intentional stance, can be expressed in quantitative terms that allow for comparative welfare assessments of different observed patterns in empirical agents' risky choices. Thus they promote what might be called a Dennettian normative empirical microeconomics. Kincaid and Nefdt argue, based on both general approaches to causal modeling in social and behavioral science and application to linguistic theory, that inferences based on real-pattern identification must be supplemented by distinct causal analyses—that is, by further efforts to isolate causal patterns from others—to support the most important kind of scientific inference.

Finally, Cao brings the menu of topics full circle by returning to Dennett's original motivating issues around the reality of different hypothesized cognitive representations, but now in the updated context of the large language models (LLMs) that are currently roiling the domains of social communication, education, corporate finance, and markets for human skills. In the last years of his life, Dennett was much preoccupied by threats to social and moral stability that he saw in underregulated propagation of LLMs. The patterns they exploit, Dennett contended, were indeed real but crucially different from those that support the integrity of human interactions and prospects for upholding crucial concepts of moral responsibility. He advocated criminalizing all attempts to publicly produce "counterfeit people," just as all countries sanction the circulation of counterfeit money. But to what extent can we identify clear lines between real human cognitive patterns and real patterns that merely imitate them? Cao explores the subtlety and complexity of efforts to compare LLMs with natural cognitive equipment, taking care to avoid a stronger conclusion than the facts can support. Of all of the chapters in the book, this is the one we most wish Dennett had had time to read and offer reflections on. His comments, we speculate, might have furnished the most direct coda to "Real Patterns" itself.

But nature has been cruel. Starting from Cao's insightful exploration, along with the other chapters here, it remains to all of us who are inspired by Dennett's ideas to try to produce that coda collectively.

References

Andersen, H. K. 2017. "Patterns, Information, and Causation." *Journal of Philosophy* 114 (11): 592–622.

Anscombe, G.E.M. 1957. *Intention.* Basil Blackwell.

Bohr, N. 1913. "I. On the Constitution of Atoms and Molecules." *London, Edinburgh, and Dublin Philosophical Magazine and Journal of Science* 26 (151): 1–25.

Bohr, N. 1923. "The Structure of the Atom." *Nature* 112 (2801): 29–44.

Bolhuis, J. J., I. Tattersall, N. Chomsky, and R. C. Berwick. 2014. "How Could Language Have Evolved?" *PLoS Biology* 12 (8): e1001934.

Burnston, D. C. 2017. "Real Patterns in Biological Explanation." *Philosophy of Science* 84 (5): 879–891.

Churchland, P. M. 1981. "Eliminative Materialism and the Propositional Attitudes." *Journal of Philosophy* 78 (2): 67–90.

Dahlbom, B., ed. 1993. *Dennett and His Critics.* Blackwell.

Dennett, D. C. 1971. "Intentional Systems." *Journal of Philosophy* 68 (4): 87–106.

Dennett, D. C. 1989. *The Intentional Stance.* MIT Press.

Dennett, D. C. 1991. "Real Patterns." *Journal of Philosophy* 88 (1): 27–51.

Dennett, D. C. 2009. "Intentional Systems Theory." In *The Oxford Handbook of Philosophy of Mind,* edited by A. Beckermann, B. P. McLaughlin, and S. Walter. Oxford University Press.

Fodor, J. A. 1974. "Special Sciences (Or: The Disunity of Science as a Working Hypothesis)." *Synthese* 28 (2): 97–115.

Fodor, J. A. 1975. *The Language of Thought.* Harvard University Press.

Fraassen, B. C. van. 1980. *The Scientific Image.* Oxford University Press.

Godfrey-Smith, P. 2006. "The Strategy of Model-Based Science." *Biology and Philosophy* 21:725–740.

Grünwald, P., and P. Vitányi. 2004. "Shannon Information and Kolmogorov Complexity." *arXiv preprint cs/0410002.*

Hartley, R. V. 1928. "Transmission of Information." *Bell System Technical Journal* 7 (3): 535–563.

Haugeland, J. 1993. "Pattern and Being." In *Dennett and His Critics,* edited by B. Dahlbom, 53–69. Blackwell.

Huffman, D. A. 1952. "A Method for the Construction of Minimum-Redundancy Codes." *Proceedings of the IRE* 40 (9): 1098–1101.

Kästner, L., and P. Haueis. 2021. "Discovering Patterns: On the Norms of Mechanistic Inquiry." *Erkenntnis* 86 (6): 1635–1660.

Kinney, D. 2019. "On the Explanatory Depth and Pragmatic Value of Coarse-Grained, Probabilistic, Causal Explanations." *Philosophy of Science* 86 (1): 145–167.

Kinney, D., and T. Lombrozo. 2024. "Tell Me Your (Cognitive) Budget, and I'll Tell You What You Value." *Cognition* 247: 105782.

Ladyman, J., and D. Ross. 2007. *Every Thing Must Go.* Oxford University Press.

Ladyman, J., and D. Ross. 2013. "The World in the Data." In *Scientific Metaphysics,* edited by H. Kincaid, J. Ladyman, and D. Ross, 108–151. Oxford University Press.

Mendeleev, D. I. 1869. "On the Correlation Between the Properties of the Elements and Their Atomic Weights." *Zhurnal Russkogo Khimicheskogo Obshchestva* 1: 60–77.

Millhouse, T. 2019. "A Simplicity Criterion for Physical Computation." *British Journal for the Philosophy of Science* 70 (1): 153–178.

Millhouse, T. 2021. "Compressibility and the Reality of Patterns." *Philosophy of Science* 88 (1): 22–43.

Millhouse, T. 2022. "Really Real Patterns." *Australasian Journal of Philosophy* 100 (4): 664–678.

Newell, A. 1980. "Physical Symbol Systems." *Cognitive Science* 4 (2): 135–183.

Petersen, S. 2019. "Composition as Pattern." *Philosophical Studies* 176 (5): 1119–1139. ISSN: 1573-0883. https://doi.org/10.1007/s11098-018-1050-6.

Petersen, Steve. 2023. "Multiple Patterns, Multiple Explanations." In *Conjunctive Explanations,* edited by J. N. Schupbach and D. H. Glass, 38–48. Routledge Studies in the Philosophy of Science. Routledge.

Putnam, H. 1988. *Representation and Reality.* MIT Press.

Quine, W.V.O. 1948. "On What There Is." *Review of Metaphysics* 2 (5): 21–38. ISSN: 00346632. http://www.jstor.org/stable/20123117.

Quine, W.V.O. 1951. "Two Dogmas of Empiricism." *Philosophical Review* 60: 20–43.

Rorty, R. 1993. "Holism, Intrinsicality, and the Ambition of Transcendence." In *Dennett and His Critics,* edited by B. Dahlbom, 184–202. Blackwell.

Ross, D. 2000. "Rainforest Realism: A Dennettian Theory of Existence." In *Dennett's Philosophy: A Comprehensive Assessment,* edited by D. Ross, A. Brook, and D. Thompson, 147–168. MIT Press.

Ross, D., A. Brook, and D. Thompson, eds. 2000. *Dennett's Philosophy: A Comprehensive Assessment.* 147–168. MIT Press/Bradford.

Sellars, W. 1962. "Philosophy and the Scientific Image of Man." In *Frontiers of Science and Philosophy,* edited by R. G. Colodny, 35–78. University of Pittsburgh Press.

Storey, J. D. 2003. "The Positive False Discovery Rate: A Bayesian Interpretation and the Q-Value." *Annals of Statistics* 31 (6): 2013–2035.

Suñé, A., and M. Martínez. 2021. "Real Patterns and Indispensability." *Synthese* 198 (5): 4315–4330. https://doi.org/10.1007/s11229-019-02343-1.

Sutton, R. S., and A. G. Barto. 1998. *Reinforcement Learning: An Introduction.* MIT Press.

Wallace, D. 2012. *The Emergent Multiverse: Quantum Theory According to the Everett Interpretation.* Oxford University Press.

Wan, E. A. 1990. "Neural Network Classification: A Bayesian Interpretation." *IEEE Transactions on Neural Networks* 1 (4): 303–305.

2 Real Patterns

Daniel C. Dennett

Are there really beliefs? Or are we learning (from neuroscience and psychology, presumably) that, strictly speaking, beliefs are figments of our imagination, items in a superseded ontology? Philosophers generally regard such ontological questions as admitting just two possible answers: Either beliefs exist or they do not. There is no such state as quasi existence; there are no stable doctrines of semirealism. Beliefs must either be vindicated along with the viruses or banished along with the banshees. A bracing conviction prevails, then, to the effect that when it comes to beliefs (and other mental items) one must be either a realist or an eliminative materialist.

2.1 Realism About Beliefs

This conviction prevails in spite of my best efforts over the years to undermine it with various analogies: Are *voices* in your ontology (Dennett 1969)? Are *centers of gravity* in your ontology (Dennett 1981)?

It is amusing to note that my analogizing beliefs to centers of gravity has been attacked from both sides of the ontological dichotomy, by philosophers who think it is simply obvious that centers of gravity are useful fictions, and by philosophers who think it is simply obvious that centers of gravity are perfectly real:

> The trouble with these supposed parallels . . . is that they are all strictly speaking *false*, although they are no doubt useful simplifications for many purposes. It is false, for example, that the gravitational attraction between the Earth and the Moon involves two point masses; but it is a good enough first approximation for many calculations. However, this is not at all what Dennett really wants to say about intentional states. For he insists that to adopt the intentional stance and interpret an agent as acting on certain beliefs and desires is to discern a pattern in his actions which is genuinely there (a pattern which is missed if we instead adopt a scientific stance): Dennett certainly does not hold that the role of intentional ascriptions is merely to give us a useful approximation to a truth that can be more accurately expressed in non-intentional terms. (Smith 1988, 22)

Compare this with Fred Dretske's (1988) equally confident assertion of realism:

> I am a realist about centers of gravity . . . The earth obviously exerts a gravitational attraction on *all* parts of the moon—not just its center of gravity. The *resultant* force, a vector sum, acts through a point, but this is something quite different. One should be very clear about what centers of gravity are before deciding whether to be literal about them, *before* deciding whether or not to be a center-of-gravity realist. (Dretske 1988, 511)

Dretske's advice is well taken. What are centers of gravity? They are mathematical points—abstract objects or what Hans Reichenbach called *abstracta*—definable in terms of physical forces and other properties. The question of whether abstract objects are real—the question of whether or not "one should be a realist about them"—can take two different paths, which we might call the metaphysical and the scientific. The metaphysical path simply concerns the reality or existence of abstract objects generally and does not distinguish them in terms of their scientific utility. Consider, for instance, the *center of population* of the United States. I define this as the mathematical point at the intersection of the two lines such that there are as many inhabitants north as south of the latitude, and as many inhabitants east as west of the longitude. This point is (or can be) just as precisely defined as the center of gravity or center of mass of an object. (Since these median strips might turn out to be wide, take the midline of each strip as the line; count as inhabitants all those within the territorial waters and up to twenty miles in altitude—orbiting astronauts do not count—and take each inhabitant's navel to be the determining point, etc.) I do not know the center of population's current geographic location, but I am quite sure it is west of where it was ten years ago. It jiggles around constantly, as people move about, taking rides on planes, trains, and automobiles, etc. I doubt that this abstract object is of any value at all in any scientific theory, but just in case it is, here is an even more trivial abstract object: Dennett's lost sock center: the point defined as the center of the smallest sphere that can be inscribed around all the socks I have ever lost in my life.

These abstract objects have the same metaphysical status as centers of gravity. Is Dretske a realist about them all? Should we be? I do not intend to pursue this question, for I suspect that Dretske is—and we should be—more interested in the scientific path to realism: Centers of gravity are real because they are (somehow) *good* abstract objects. They deserve to be taken seriously, learned about, used. If we go so far as to distinguish them as *real* (contrasting them, perhaps, with those abstract objects which are *bogus*), that is because we think they serve in perspicuous representations of real forces, "natural" properties, and the like. This path brings us closer, in any case, to the issues running in the debates about the reality of beliefs.

I have claimed that beliefs are best considered to be abstract objects rather like centers of gravity. Smith considers centers of gravity to be useful fictions, while Dretske considers them to be useful (and hence?) real abstractions, and each takes his view to constitute a criticism of my position. The optimistic assessment of these opposite criticisms is that they cancel each other out; my analogy must have hit the nail on the head. The pessimistic assessment is that more needs to be said to convince philosophers that a mild and intermediate sort of realism is a positively attractive position, and not just the desperate dodge of ontological responsibility it has sometimes been taken to be. I have just such a case to present, a generalization and extension of my earlier attempts, via the concept of a *pattern*. My aim on this occasion is not so much to prove that my intermediate doctrine about the reality of psychological states is right, but just that it is quite possibly right because a parallel doctrine is demonstrably right about some simpler cases.

We use folk psychology—interpretation of each other as believers, wanters, intenders, and the like—to predict what people will do next. Prediction is not the only thing we care about, of course. Folk psychology helps us understand and empathize with others, organize our memories, interpret our emotions, and flavor our vision in a thousand ways, but at the heart of all these is the enormous predictive leverage of folk psychology. Without its predictive power, we could have no interpersonal projects or relations at all; human activity would be just so much Brownian motion; we would be baffling ciphers to each other and to ourselves—we could not even conceptualize our own flailings. In what follows, I shall concentrate always on folk-psychological prediction, not because I make the mistake of ignoring all the other interests we have in people aside from making bets on what they will do next, but because I claim that our power to *interpret* the actions of others depends on our power—seldom explicitly exercised—to predict them.[1]

1. R. A. Sharpe (1989) takes me to task on this point, using examples from Proust to drive home the point that "Proust draws our attention to possible lives and these possible lives are various. But in none of them is prediction of paramount importance" (240). I agree. I also agree that what makes people interesting (in novels and in real life) is precisely their unpredictability. But that unpredictability is only interesting against the backdrop of routine predictability on which all interpretation depends. As I note in *The Intentional Stance* (Dennett 1989, 79) in response to a similar objection of Fodor's, the same is true of chess: The game is interesting only because of the unpredictability of one's opponent, but that is to say: The intentional stance can usually eliminate only ninety percent of the legal moves.

Where utter patternlessness or randomness prevails, nothing is predictable. The success of folk-psychological prediction, like the success of any prediction, depends on there being some order or pattern in the world to exploit. Exactly where in the world does this pattern exist? What is the pattern a pattern *of* (Nelkin 1994)? Some have thought, with Fodor, that the pattern of belief must in the end be a pattern of structures in the brain, formulae written in the language of thought. Where else could it be? Gibsonians might say the pattern is "in the light"—and Quinians (such as Donald Davidson and I) could almost agree: The pattern is discernible in agents' (observable) behavior when we subject it to "radical interpretation" (Davidson) "from the intentional stance" (Dennett).

When are the elements of a pattern real and not merely apparent? Answering this question will help us resolve the misconceptions that have led to the proliferation of "ontological positions" about beliefs, the different grades or kinds of realism. I shall concentrate on five salient exemplars arrayed in the space of possibilities: Fodor's industrial-strength Realism (he writes it with a capital "R"); Davidson's regular strength realism; my mild realism; Richard Rorty's milder-than-mild irrealism, according to which the pattern is only in the eyes of the beholders; and Paul Churchland's eliminative materialism, which denies the reality of beliefs altogether.

In what follows, I shall assume that these disagreements all take place within an arena of common acceptance of what Arthur Fine (1986) calls NOA, the natural ontological attitude.[2] That is, I take the interest in these disagreements to lie not in differences of opinion about the ultimate metaphysical status of physical things or abstract things (e.g., electrons or centers of gravity), but in differences of opinion about whether beliefs and other mental states are, shall we say, *as real as* electrons or centers of gravity. I want to show that mild realism is the doctrine that makes the most sense when what we are talking about is real patterns, such as the real patterns discernible from the intentional stance.[3]

In order to make clear the attractions and difficulties of these different positions about patterns, I shall apply them first to a much simpler, more readily visualized, and uncontroversial sort of pattern.

2. See especially p. 153n and his comments there on Rorty, which I take to be consonant with mine here.

3. See *The Intentional Stance*, "Real Patterns, Deeper Facts, and Empty Questions" (Dennett 1989, 38–42).

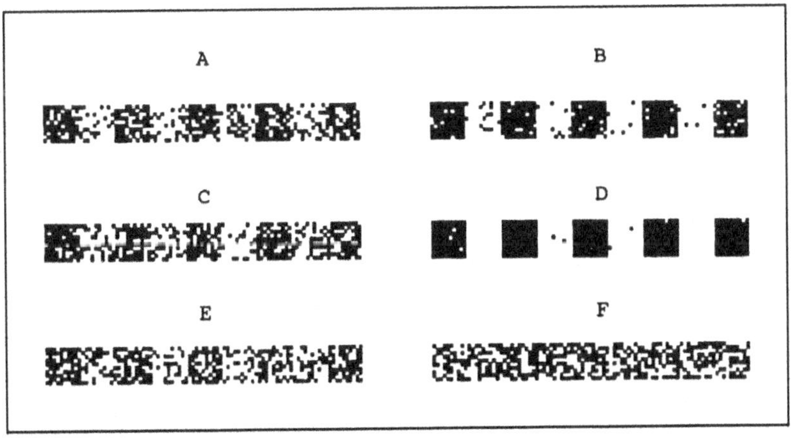

Figure 2.1

2.2 The Reality of Patterns

Consider the six objects in figure 2.1 (which I shall call frames). We can understand a frame to be a finite subset of data, a window on an indefinitely larger world of further data. In one sense *A–F* all display different patterns; if you look closely you will see that no two frames are exactly alike ("atom-for-atom replicas," if you like). In another sense, *A–F* all display the same pattern; they were all made by the same basic process: a printing of ten rows of ninety dots, ten black dots followed by ten white dots, etc. The overall effect is to create five equally spaced black squares or bars in the window. I take it that this pattern, which I shall dub *bar code*, is a real pattern if anything is. But some random (actually pseudo-random) "noise" has been allowed to interfere with the actual printing. The noise ratio is as follows:

A: 25%	*B*: 10%
C: 25%	*D*: 1%
E: 33%	*F*: 50%

It is impossible to see that *F* is not purely (pseudo-) random noise; you will just have to take my word for it that it was actually generated by the same program that generated the other five patterns; all I changed was the noise ratio.

Now, what does it mean to say that a pattern in one of these frames is real, or that it is really there? Given our privileged information about how these frames were generated, we may be tempted to say that there is a single pattern in all six cases—even in F, where it is "indiscernible." But I propose that the self-contradictory air of "indiscernible pattern" should be taken seriously. We may be able to make some extended, or metaphorical, sense of the idea of indiscernible patterns (or invisible pictures or silent symphonies), but in the root case a pattern is "by definition" a candidate for pattern *recognition.* (It is this loose but unbreakable link to observers or perspectives, of course, that makes "pattern" an attractive term to someone perched between instrumentalism and industrial-strength realism.)

Fortunately, there is a standard way of making these intuitions about the discernibility-in-principle of patterns precise. Consider the task of transmitting information about one of the frames from one place to another. How many bits of information will it take to transmit each frame? The least efficient method is simply to send the "bit map," which identifies each dot *seriatim* ("dot one is black, dot two is white, dot three is white . . ."). For a black-and-white frame of 900 dots (or pixels, as they are called), the transmission requires 900 bits. Sending the bit map is in effect verbatim quotation, accurate but inefficient. Its most important virtue is that it is equally capable of transmitting any pattern or any particular instance of utter patternlessness.

Gregory Chaitin's (1975) valuable definition of mathematical randomness invokes this idea. A series (of dots or numbers or whatever) is random if and only if the information required to describe (transmit) the series accurately is incompressible: Nothing shorter than the verbatim bit map will preserve the series. Then a series is not random—has a pattern—if and only if there is some more efficient way of describing it.[4] Frame D, for instance, can be described as "ten rows of ninety, ten black followed by ten white, etc., *with the following exceptions*: dots 57, 88 . . ." This expression, suitably encoded, is much shorter than 900 bits long. The comparable expressions for the other frames will be proportionally longer, since they will have to mention, verbatim, more exceptions, and the degeneracy of the "pattern" in F is revealed by the fact that its description in this system will be no improvement over the bit map—in fact, it will tend on average to be trivially longer, since it takes some bits to describe the pattern that is then obliterated by all the exceptions.

Of course, there are bound to be other ways of describing the evident patterns in these frames, and some will be more efficient than others—in the

4. More precisely: "A series of numbers is random if the smallest algorithm capable of specifying it to a computer has about the same number of bits of information as the series itself" (Chaitin, 48). This is what explains the fact that the "random number generator" built into most computers is not really properly named, since it is some function describable in a few bits (a little subroutine that is called for some output whenever a program requires a "random" number or series). If I send you the description of the pseudo-random number generator on my computer, you can use it to generate exactly the same infinite series of random seeming digits.

precise sense of being systematically specifiable in fewer bits.[5] Any such description, if an improvement over the bit map, is the description of a real pattern in the data.[6]

Consider bar code, the particular pattern seen in *A–E*, and almost perfectly instantiated in *D*. *That* pattern is quite readily discernible to the naked human eye in these presentations of the data because of the particular pattern-recognition machinery hard-wired in our visual systems—edge detectors, luminance detectors, and the like. But the very same data (the very same streams of bits) presented in some other format might well yield no hint of pattern to us, especially in the cases where bar code is contaminated by salt and pepper, as in frames *A* through *C*. For instance, if we broke the 900-bit series of frame *B* into 4-bit chunks and then translated each of these into hexadecimal notation, one would be hard pressed indeed to tell the resulting series of hexadecimal digits from a random series, since the hexadecimal chunking would be seriously out of phase with the decimal pattern—and hence the "noise" would not "stand out" as noise. There are myriad ways of displaying any 900-bit series of data points, and not many of them would inspire us to concoct an efficient description of the series. Other creatures with different sense organs, or different interests, might readily perceive patterns that were imperceptible to us. The patterns would be *there* all along, but just invisible to *us*.

The idiosyncracy of perceivers' capacities to discern patterns is striking. Visual patterns with axes of vertical symmetry stick out like sore thumbs for us, but if one simply rotates the frame a few degrees, the symmetry is often utterly beyond noticing. And the "perspectives" from which patterns are "perceptible" are not restricted to variations on presentation to the sense modalities. Differences in knowledge yield striking differences in the capacity to pick up patterns. Expert chess players can instantly perceive (and subsequently recall with high accuracy) the total board position in a real game but are much worse at recall if the same chess pieces are randomly placed on the board, even though

5. Such schemes for efficient description, called compression algorithms, are widely used in computer graphics for saving storage space. They break the screen into uniformly colored regions, for instance, and specify region boundaries (rather like the "paint by numbers" line drawings sold in craft shops). The more complicated the picture on the screen, the longer the compressed description will be; in the worst case (a picture of confetti randomly sprinkled over the screen) the compression algorithm will be stumped and can do no better than a verbatim bit map.

6. What about the "system" of pattern description that simply baptizes frames with proper names (*A* through *F*, in this case) and tells the receiver which frame is up by simply sending "F"? This looks much shorter than the bit map until we consider that such a description must be part of an entirely general system. How many proper names will we need to name all possible 900-dot frames? Trivially, the 900-bit binary number, 11111111 . . . To send the "worst-case" proper name will take exactly as many bits as sending the bit map. This confirms our intuition that proper names are maximally inefficient ways of couching generalizations ("Alf is tall and Bill is tall and . . .").

to a novice both boards are equally hard to recall (de Groot 1965). This should not surprise anyone who considers that an expert speaker of English would have much less difficulty perceiving and recalling

> The frightened cat struggled to get loose.

than

> Te serioghehnde t srugfcalde go tgtt ohle

which contains the same pieces, now somewhat disordered. Expert chess players, unlike novices, not only know how to Play chess; they know how to read chess—how to see the patterns at a glance.

A pattern exists in some data—is real—if *there is* a description of the data that is more efficient than the bit map, whether or not anyone can concoct it. Compression algorithms, as general-purpose pattern describers, are efficient ways of transmitting exact copies of frames, such as *A–F*, from one place to another, but our interests often favor a somewhat different goal: transmitting *inexact* copies that nevertheless preserve "the" pattern that is important to us. For some purposes, we need not list the exceptions to bar code, but only transmit the information that the pattern is bar code with *n%* noise. Following this strategy, frames A and C, though discernibly different under careful inspection, count as *the same pattern*, since what matters to us is that the pattern is bar code with 25% noise, and we do not care which particular noise occurs, only that it occurs.

Sometimes we are interested in not just ignoring the noise, but eliminating it, improving the pattern in transmission. Copy-editing is a good example. Consider the likely effect thes santince wull hive hod on tha cupy adutor whu preparis thas monescrupt fur prunteng. *My* interest in this particular instance is that the "noise" be transmitted, not removed, though I actually do not care exactly *which* noise is there.

Here then are three different attitudes we take at various times toward patterns. Sometimes we care about exact description or reproduction of detail, at whatever cost. From this perspective, a real pattern in frame *A* is bar code with the following exceptions: 7, 8, 11 . . . At other times we care about the noise, but not where in particular it occurs. From this perspective, a real pattern in frame *A* is *bar code with 25% noise*. And sometimes, we simply tolerate or ignore the noise. From this perspective, a real pattern in frame *A* is simply: *bar code*. But is bar code really there in frame *A*? I am tempted to respond: Look! You can see it with your own eyes. But there is something more constructive to say as well.

When two individuals confront the same data, they may perceive different patterns in them, but since we can have varied interests and perspectives, these differences do not all count as disagreements. Or in any event they should not. If Jones sees pattern α (with $n\%$ noise) and Brown sees pattern β (with $m\%$ noise), there may be no ground for determining that one of them is right and the other wrong. Suppose they are both using their patterns to bet on the next datum in the series. Jones bets according to the "pure" pattern α, but budgets for $n\%$ errors when he looks for odds. Brown does likewise, using pattern β. If both patterns are real, they will both get rich. That is to say, so long as they use their expectation of deviations from the "ideal" to temper their odds policy, they will do better than chance—perhaps very much better.

Now suppose they compare notes. Suppose that α is a simple, easy-to-calculate pattern, but with a high noise rate—for instance, suppose α is bar code as it appears in frame E. And suppose that Brown has found some periodicity or progression in the "random" noise that Jones just tolerates, so that β is a much more complicated description of pattern-superimposed-on-pattern. This permits Brown to do better than chance, we may suppose, at predicting when the "noise" will come. As a result, Brown budgets for a lower error rate—say only 5%. "What you call noise, Jones, is actually pattern," Brown might say. "Of course there is still *some* noise in my pattern, but my pattern is better—more real—than yours! Yours is actually just a mere appearance." Jones might well reply that it is all a matter of taste; he notes how hard Brown has to work to calculate predictions and points to the fact that he is getting just as rich (or maybe richer) by using a simpler, sloppier system and making more bets at good odds than Brown can muster. "My pattern is perfectly real—look how rich I'm getting. If it were an illusion, I'd be broke."

This crass way of putting things—in terms of betting and getting rich—is simply a vivid way of drawing attention to a real, and far from crass, trade-off that is ubiquitous in nature, and hence in folk psychology. Would we prefer an extremely compact pattern description with a high noise ratio or a less compact pattern description with a lower noise ratio? Our decision may depend on how swiftly and reliably we can discern the simple pattern, how dangerous errors are, how much of our resources we can afford to allocate to detection and calculation. These "design decisions" are typically not left to us to make by individual and deliberate choices; they are incorporated into the design of our sense organs by genetic evolution, and into our culture by cultural evolution. The product of this design evolution process is what Wilfrid Sellars (1963) calls our *manifest image*, and it is composed of folk physics, folk psychology, and the other pattern-making perspectives we have on the buzzing blooming confusion that bombards us with data. The ontology generated by the manifest image has thus a deeply pragmatic source.[7]

Do these same pragmatic considerations apply to the scientific image, widely regarded as the final arbiter of ontology? Science is supposed to carve nature at the joints—at its *real* joints, of course. Is it permissible in science to adopt a carving system so simple that it makes sense to tolerate occasional misdivisions and consequent mispredictions? It happens all the time. The ubiquitous practice of using idealized models is exactly a matter of trading off reliability and accuracy of prediction against computational tractability. A particularly elegant and handy oversimplification may under some circumstances be irresistible. The use of Newtonian rather than Einsteinian mechanics in most mundane scientific and engineering calculations is an obvious example. A tractable oversimplification may be attractive even in the face of a high error rate; considering inherited traits to be carried by single genes "for" those traits is an example; considering agents in the marketplace to be perfectly rational self-aggrandizers with perfect information is another.

2.3 Patterns in Life

The time has come to export these observations about patterns and reality to the controversial arena of belief attribution. The largish leap we must make is nicely expedited by pausing at a stepping-stone example midway between the world of the dot frames and the world of folk psychology: John Horton Conway's Game of Life. In my opinion, every philosophy student should be held responsible for an intimate acquaintance with the Game of Life. It should be considered an essential tool in every thought-experimenter's kit, a prodigiously versatile generator of philosophically important examples and thought experiments of admirable clarity and vividness. In *The Intentional Stance* (Dennett 1989, 37–39), I briefly exploited it to make a point about the costs and benefits of risky prediction from the intentional stance, but I have since learned that I presumed too much familiarity with the underlying ideas. Here, then, is a somewhat expanded basic introduction to Life.[8]

7. In "Randomness and Perceived Randomness in Evolutionary Biology," William Wimsatt (1980) offers a nice example (296): While the insectivorous bird tracks individual insects, the anteater just averages over the ant-infested area; one might say that, while the bird's manifest image quantifies over insects, "ant" is a mass term for anteaters. See the discussion of this and related examples in my *Elbow Room* (Dennett 1984, 108–110).

8. Martin Gardner introduced the Game of Life to a wide audience in two columns in *Scientific American* in October 1970 and February 1971. William Poundstone, *The Recursive Universe: Cosmic Complexity and the Limits of Scientific Knowledge* (1985), is an excellent exploration of the game and its philosophical implications. Two figures from Poundstone's book are reproduced,

Life is played on a two-dimensional grid, such as a checkerboard or a computer screen; it is not a game one plays to win; if it is a game at all, it is solitaire. The grid divides space into square cells, and each cell is either ON or OFF at each moment. Each cell has eight neighbors: the four adjacent cells north, south, east, and west, and the four diagonals: northeast, southeast, southwest, and northwest. Time in the Life world is also discrete, not continuous; it advances in ticks, and the state of the world changes between each tick according to the following rule:

> Each cell, in order to determine what to do in the next instant, counts how many of its eight neighbors is ON at the present instant. If the answer is exactly two, the cell stays in its present state (ON or OFF) in the next instant. If the answer is exactly three, the cell is ON in the next instant whatever its current state. Under all other conditions the cell is OFF.

The entire physics of the Life world is captured in that single, unexceptioned law. (While this is the fundamental law of the "physics" of the Life world, it helps at first to conceive this curious physics in biological terms: Think of cells going ON as births, cells going OFF as deaths, and succeeding instants as generations. Either overcrowding [more than three inhabited neighbors] or isolation [less than two inhabited neighbors] leads to death.) By the scrupulous application of this single law, one can predict with perfect accuracy the next instant of any configuration of ON and OFF cells, and the instant after that, and so forth. In other words, the Life world is a toy world that perfectly instantiates Laplace's vision of determinism: Given the state description of this world at an instant, we finite observers can perfectly predict the future instants by the simple application of our one law of physics. Or, in my terms, when we adopt the physical stance toward a configuration in the Life world, our powers of prediction are perfect: There is no noise, no uncertainty, no probability less than one. Moreover, it follows from the two-dimensionality of the Life world that nothing is hidden from view. There is no backstage; there are no hidden variables; the unfolding of the physics of objects in the Life world is directly and completely visible.

There are computer simulations of the Life world in which one can set up configurations on the screen and then watch them evolve according to the single rule. In the best simulations, one can change the scale of both time and space, alternating between close-up and bird's-eye view. A nice touch added to some color versions is that ON cells (often just called pixels) are color-coded

---

with kind permission from the author and publisher. (Note from the editors: We have reconstructed these figures in this printing to improve readability. The content of the figures is unchanged.)

by their age; they are born blue, let us say, and then change color each generation, moving through green to yellow to orange to red to brown to black and then staying black unless they die. This permits one to see at a glance how old certain patterns are, which cells are co-generational, where the birth action is, and so forth.[9]

One soon discovers that some simple configurations are more interesting than others. In addition to those configurations which never change—the "still lifes" such as four pixels in a square—and those which evaporate entirely—such as any long diagonal line segment, whose two tail pixels die of isolation each instant until the line disappears entirely—there are configurations with all manner of periodicity. Three pixels in a line make a simple flasher, which becomes three pixels in a column in the next instant, and reverts to three in a line in the next, ad infinitum, unless some other configuration encroaches. Encroachment is what makes Life interesting: Among the periodic configurations are some that swim, amoeba-like, across the plane. The simplest is the *glider*, the five-pixel configuration shown taking a single stroke to the southeast in figure 2.2. Then there are the eaters, the puffer trains, and space rakes, and a host of other aptly named denizens of the Life world that emerge in the ontology of a new level, analogous to what I have called the design level. This level has its own language, a transparent foreshortening of the tedious descriptions one could give at the physical level. For instance:

> An eater can eat a glider in four generations. Whatever is being consumed, the basic process is the same. A bridge forms between the eater and its prey. In the next generation, the bridge region dies from overpopulation, taking a bite out of both eater and prey. The eater then repairs itself. The prey usually cannot. If the remainder of the prey dies out as with the glider, the prey is consumed. (Poundstone 1985, 38)

Note that there has been a distinct ontological shift as we move between levels; whereas at the physical level there is no motion, and the only individuals, cells, are defined by their fixed spatial location, at this design level we have the motion of persisting objects; it is one and the same glider that has moved southeast in figure 2.2, changing shape as it moves, and there is one less glider in the world after the eater has eaten it in figure 2.3. (Here is a warming-up exercise for what is to follow: Should we say that there is *real* motion in the Life world, or only *apparent* motion? The flashing pixels on the computer screen are a paradigm case, after all, of what a psychologist would call apparent motion. Are there *really* gliders that move, or are there just patterns of cell state that

9. Poundstone (1985) provides simple BASIC and IBM-PC assembly language simulations you can copy for your own home computer, and describes some of the interesting variations.

move? And if we opt for the latter, should we say at least that these moving patterns are real?)

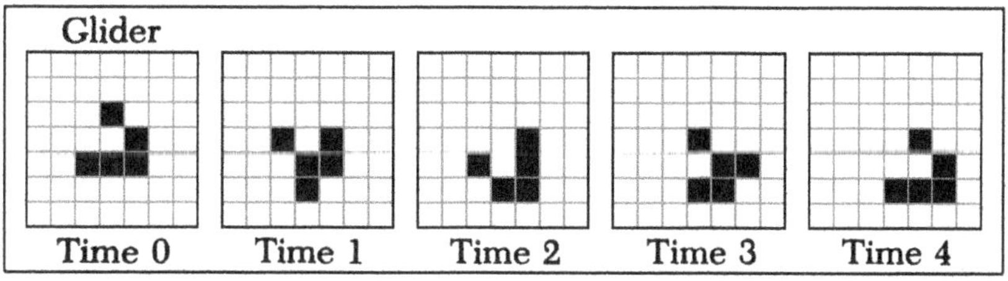

Figure 2.2: Poundstone (1985).

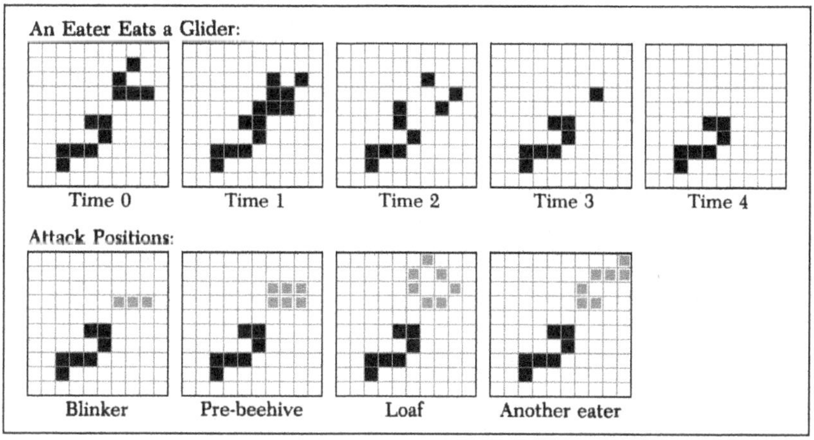

Figure 2.3: Poundstone (1985).

Notice, too, that at this level one proposes generalizations that require "usually" or "provided nothing encroaches" clauses. Stray bits of debris from earlier events can "break" or "kill" one of the objects in the ontology at this level; their *salience as real things* is considerable, but not guaranteed. To say that their salience is considerable is to say that one can, with some small risk, ascend to this design level, adopt its ontology, and proceed to predict—sketchily and riskily—the behavior of larger configurations or systems of configurations, without bothering to compute the physical level. For instance, one can set oneself the task of designing some interesting supersystem out of the "parts" that the design level makes available. Surely the most impressive triumph of this design activity in the Life world is the proof that a working model of a universal Turing machine can in principle be constructed in the Life plane! Von Neumann had already shown that in principle a two-dimensional universal Turing machine could be constructed out of cellular automata, so it wasn't "just" a matter of "engineering" to show how, in principle, it could be constructed out of the simpler cellular automata defined in the Life world. Glider streams can provide the tape, for instance, and the tape reader can be some huge assembly of eaters, gliders, and other bits and pieces. What does this huge Turing machine look like? Poundstone calculates that the whole construction, a self-reproducing machine incorporating a universal Turing machine, would be on the order of 10^{13} pixels.

> Displaying a 10^{13}-pixel pattern would require a video screen about 3 million pixels across at least. Assume the pixels are 1 millimeter square (which is very high resolution by the standards of home computers). Then the screen would have to be 3 kilometers (about two miles) across. It would have an area about six times that of Monaco. Perspective would shrink the pixels of a self-reproducing pattern to invisibility. If you got far enough away from the screen so that the entire pattern was comfortably in view, the pixels (and even the gliders, eaters and guns) would be too tiny to make out. A self-reproducing pattern would be a hazy glow, like a galaxy. (Poundstone 1985, 227–228)

Now, since the universal Turing machine can compute any computable function, it can play chess—simply by mimicking the program of any chess-playing computer you like. Suppose, then, that such an entity occupies the Life plane, playing chess against itself. Looking at the configuration of dots that accomplishes this marvel would almost certainly be unilluminating to anyone who had no clue that a configuration with such powers could exist. But from the perspective of one who had the hypothesis that this huge array of black dots was a chess-playing computer, enormously efficient ways of predicting the future of that configuration are made available. As a first step one can shift from an ontology of gliders and eaters to an ontology of symbols and machine states, and, adopting this higher design stance toward the configuration, predict its future *as* a Turing machine. As a second and still more efficient step, one can shift to an ontology of chess-board positions, possible chess moves, and the grounds for evaluating them; then, adopting the intentional stance toward the configuration, one can predict its future as a chess player performing intentional actions—making chess moves and trying to achieve checkmate. Once one has fixed on an interpretation scheme, permitting one to say which configurations of pixels count as which symbols (either, at the Turing machine level, the symbols "0" or "1", say, or at the intentional level, "*QxBch*" and the other symbols for chess moves), one can use the interpretation scheme to predict, for instance, that the next configuration to emerge from the galaxy will be such-and-such a glider stream (the symbols for "*RxQ*", say). There is risk involved in either case, because the chess program being run on the Turing machine may be far from perfectly rational, and, at a different level, debris may wander onto the scene and "break" the Turing machine configuration before it finishes the game.

In other words, real but (potentially) noisy patterns abound in such a configuration of the Life world, there for the picking up if only we are lucky or clever enough to hit on the right perspective. They are not visual patterns but, one might say, intellectual patterns. Squinting or twisting the page is not apt to help, while posing fanciful interpretations (or what W. V. Quine would call "analytical hypotheses") may uncover a goldmine. The opportunity confronting the observer of such a Life world is analogous to the opportunity confronting the cryptographer staring at a new patch of cipher text, or the opportunity confronting the Martian, peering through a telescope at the Superbowl Game. If the Martian hits on the intentional stance—or folk psychology—as the right level to look for pattern, shapes will readily emerge through the noise.

2.4 The Reality of Intentional Patterns

The scale of compression when one adopts the intentional stance toward the two-dimensional chess-playing computer galaxy is stupendous: It is the difference between figuring out in your head what white's most likely (best) move is versus calculating the state of a few trillion pixels through a few hundred thousand generations. But the scale of the savings is really no greater in the Life world than in our own. Predicting that someone will duck if you throw a brick at him is easy from the folk-psychological stance; it is and will always be intractable if you have to trace the photons from brick to eyeball, the neurotransmitters from optic nerve to motor nerve, and so forth.

For such vast computational leverage one might be prepared to pay quite a steep price in errors, but in fact one belief that is shared by all of the representatives on the spectrum I am discussing is that "folk psychology" provides a description system that permits highly reliable prediction of human (and much nonhuman) behavior.[10] They differ in the explanations they offer of this predictive prowess, and the implications they see in it about "realism."

For Fodor, an industrial-strength Realist, beliefs and their kin would not be real unless the pattern dimly discernible from the perspective of folk psychology could also be discerned (more clearly, with less noise) as a pattern

10. To see that the opposite poles share this view, see Fodor (1987, chap. 1) and Paul Churchland (1979): "For the P-theory [folk psychology] is in fact a marvelous intellectual achievement. It gives its possessor an explicit and systematic insight into the behaviour, verbal and otherwise, of some of the most complex agents in the environment, and its overall prowess in that respect remains unsurpassed by anything else our considerable theoretical efforts have produced" (100).

of structures in the brain. The pattern would have to be discernible from the different perspective provided by a properly tuned *syntactoscope* aimed at the purely formal (nonsemantic) features of Mentalese terms written in the brain. For Fodor, the pattern seen through the noise by everyday folk psychologists would tell us nothing about reality, unless it, and the noise, had the following sort of explanation: What we discern from the perspective of folk psychology is the net effect of two processes: an ulterior, hidden process wherein the pattern exists quite pure, overlaid, and partially obscured by various intervening sources of noise: performance errors, observation errors, and other more or less random obstructions. He might add that the interior belief-producing process was in this respect just like the process responsible for the creation of frames *A–F*. If you were permitted to peer behind the scenes at the program I devised to create the frames, you would see, clear as a bell, the perfect bar-code periodicity, with the noise thrown on afterward like so much salt and pepper.

This is often the explanation for the look of a dataset in science, and Fodor may think that it is either the only explanation that can ever be given, or at any rate the only one that makes any sense of the success of folk psychology. But the rest of us disagree. As G.E.M. Anscombe (1957) put it in her pioneering exploration of intentional explanation, "If Aristotle's account [of reasoning using the practical syllogism] were supposed to describe actual mental processes, it would in general be quite absurd. The interest of the account is that it describes an order which is there whenever actions are done with intentions" (80).

But how could the *order* be there, so visible amidst the noise, if it were not the direct outline of a concrete orderly process in the background? Well, it *could* be there thanks to the statistical effect of very many concrete minutiae producing, as if by a hidden hand, an approximation of the "ideal" order. Philosophers have tended to ignore a variety of regularity intermediate between the regularities of planets and other objects "obeying" the laws of physics and the regularities of rule-following (that is, rule-*consulting*) systems.[11] These intermediate regularities are those which are preserved under selection pressure: the regularities dictated by principles of good design and hence homed in on by self-designing systems. That is, a "rule of thought" may be much more than a mere regularity; it may be a *wise* rule, a rule one would design a system by if one were a system designer, and hence a rule one would expect

11. A notable early exception is Sellars, who discussed the importance of just this sort of regularity in "Some Reflections on Language Games" (1954). See especially the subsection of this classic paper, entitled "Pattern Governed and Rule Obeying Behavior," reprinted in Sellars's *Science, Perception and Reality* (1963).

self-designing systems to "discover" in the course of settling into their patterns of activity. Such rules no more need be explicitly represented than do the principles of aerodynamics that are honored in the design of birds' wings.[12]

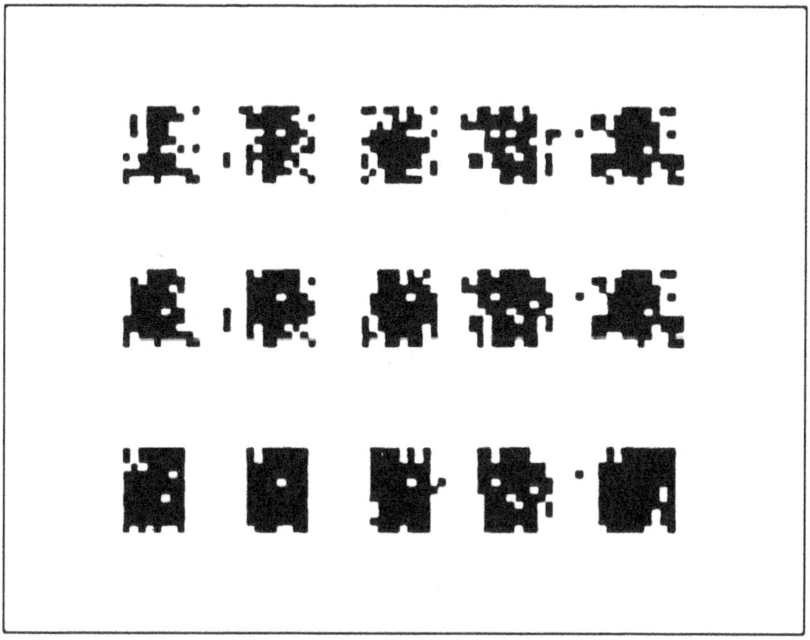

Figure 2.4

12. Several interpreters of a draft of this article have supposed that the conclusion I am urging here is that beliefs (or their contents) are epiphenomena having no causal powers, but this is a misinterpretation traceable to a simplistic notion of causation. If one finds a predictive pattern of the sort just described one has *ipso facto* discovered a causal power—a difference in the world that makes a subsequent difference testable by standard empirical methods of variable manipulation. Consider the crowd-drawing power of a sign reading "Free Lunch" placed in the window of a restaurant, and compare its power in a restaurant in New York to its power in a restaurant in Tokyo. The intentional level is obviously the right level at which to predict and explain such causal powers; the sign more reliably produces a particular belief in one population of perceivers than in the other, and variations in the color of typography of the sign are not as predictive of variations in crowd-drawing power as are variations in (perceivable) meaning. The fact that the regularities on which these successful predictions are based are efficiently capturable (only) in intentional terms and are not derived from "covering laws" does not show that the regularities are not "causal"; it just shows that philosophers have often relied on pinched notions of causality derived from exclusive attention to a few examples drawn from physics and chemistry. Smith has pointed out to me that here I am echoing Aristotle's claim that his predecessors had ignored final causes.

The contrast between these different sorts of pattern-generation processes can be illustrated. The frames in figure 2.1 were created by a hard-edged process (ten black, ten white, ten black . . .) obscured by noise, while the frames in figure 2.4 were created by a process almost the reverse of that: The top frame shows a pattern created by a normal distribution of black dots around means at x = 10, 30, 50, 70, and 90 (rather like Mach bands or interference fringes); the middle and bottom frames were created by successive applications of a very simple contrast enhancer applied to the top frame: A vertical slit "window" three pixels high is thrown randomly onto the frame; the pixels in the window vote, and majority rules. This gradually removes the salt from the pepper and the pepper from the salt, creating "artifact" edges such as those discernible in the bottom frame. The effect would be more striking at a finer pixel scale, where the black merges imperceptibly through grays to white, but I chose to keep the scale at the ten-pixel period of bar code. I do not mean to suggest that it is impossible to tell the patterns in figure 2.4 from the patterns in figure 2.1. Of course it is possible; for one thing, the process that produced the frames in figure 2.1 will almost always show edges at exactly 10, 20, 30 . . . and almost never at 9, 11, 19, 21 . . . while there is a higher probability of these "displaced" edges being created by the process of figure 2.4 (as a close inspection of figure 2.4 reveals). Fine-tuning could of course reduce these probabilities, but that is not my point. My point is that *even if* the evidence is substantial that the discernible pattern is produced by one process rather than another, it can be rational to ignore those differences and use the simplest pattern description (e.g., *bar code*) as one's way of organizing the data.

Fodor and others have claimed that an interior language of thought is the best explanation of the hard edges visible in "propositional attitude psychology." Churchland and I have offered an alternative explanation of these edges, an explanation for which the process that produced the frames in figure 2.4 is a fine visual metaphor. The process that produces the data of folk psychology, we claim, is one in which the multidimensional complexities of the underlying processes are projected *through linguistic behavior*, which creates an appearance of definiteness and precision, thanks to the discreteness of words.[13] As Churchland (1981) puts it, a person's declarative utterance is a "one-dimensional *projection*—through the compound lens of Wernicke's and

13. See my discussion of the distinction between beliefs and (linguistically infected) opinions; in *Brainstorms* (1978, chap. 16), and in "The Illusions of Realism," in *The Intentional Stance* (1989, 110–116).

Broca's areas onto the idiosyncratic surface of the speaker's language—a one-dimensional projection of a four- or five-dimensional 'solid' that is an element in his true kinematic state" (85).

Fodor's industrial-strength Realism takes beliefs to be things in the head—just like cells and blood vessels and viruses. Davidson and I both like Churchland's alternative idea of propositional-attitude statements as indirect "measurements" of a reality diffused in the behavioral dispositions of the brain (and body).[14] We think beliefs are quite real enough to call real just so long as belief talk measures these complex behavior-disposing organs as predictively as it does. What do we disagree about? As John Haugeland[15] has pointed out, Davidson is more of a realist than I am, and I have recently tracked down the source of this disagreement to a difference of opinion we have about the status of Quine's principle of indeterminacy of translation, which we both accept.

For Davidson, the principle is not the shocker it is often taken to be; in fact, it is well-nigh trivial—the two different translation manuals between which no fact of the matter decides are like two different scales for measuring temperature.

> We know there is no contradiction between the temperature of the air being 32° fahrenheit and 0° celsius; there is nothing in this "relativism" to show that the properties being measured are not "real." Curiously, though, this conclusion has repeatedly been drawn . . . Yet in the light of the considerations put forward here, this comes to no more than the recognition that more than one set of one person's utterances might be equally successful in capturing the contents of someone else's thoughts or speech. Just as numbers can capture all the empirically significant relations among weights or temperatures in infinitely many ways, so one person's utterances can capture all the significant features of another person's thoughts and speech in different ways. This fact does not challenge the "reality" of the attitudes or meanings thus reported. (Davidson 1989, 10)

On Davidson's view, no substantive disagreements emerge from a comparison of the two description schemes, and so they can quite properly be viewed as competing descriptions of the same reality.

I think this is a flawed analogy. A better one is provided by the example of "rival" descriptions of patterns-with-noise. Consider two rival intentional

14. Churchland introduces the idea in *Scientific Realism and the Plasticity of Mind* (1979, 100–107). My adoption of the idea was in "Beyond Belief," in Woodfield (1982), reprinted as chapter 5 of *The Intentional Stance* (Dennett 1989). Davidson's guarded approval is expressed in "What is Present to the Mind?" read at the Sociedad Filosófica Ibero Americana meeting in Buenos Aires, 1989. (Note from the editors: Davidson's conference paper, referenced by Dennett here, differs slightly from the version published by Davidson in 1991.)

15. See the discussion of Haugeland's views in the last chapter of *The Intentional Stance*, "Mid-Term Examination: Compare and Contrast" (Dennett 1991, 348–349).

interpretations of a single individual; they agree on the general shape of this individual's collection of beliefs (and desires, etc.), but because of their different idealizations of the pattern, they do not agree point for point. Recalling a famous analogy of Quine's[16] and extending it beyond radical translation to radical interpretation (as Davidson and I both wish to do), we get the image in figure 2.5.

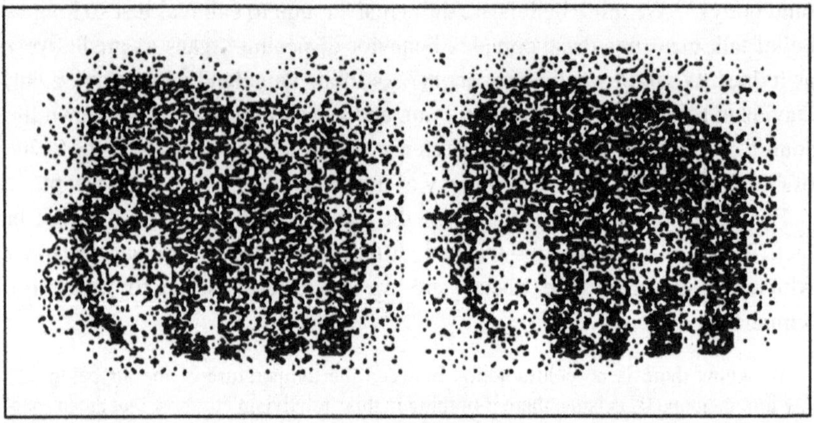

Figure 2.5

To the left we see Brown's intentional interpretation of Ella; to the right, Jones's interpretation. Since these are intentional interpretations, the pixels or data points represent beliefs and so forth, not (for instance) bits of bodily motion or organs or cells or atoms, and since these are rival intentional interpretations of a single individual, the patterns discerned are not statistical averages (e.g., "Democrats tend to favor welfare programs") but personal cognitive idiosyncracies (e.g., "She thinks she should get her queen out early"). Some of the patterns may indeed be simple observed periodicities (e.g., "Ella wants to talk about football on Mondays") but we are to understand the pattern to be what Anscombe called the "order which is there" in the rational coherence of a person's set of beliefs, desires, and intentions.

Notice that here the disagreements can be substantial—at least before the fact: When Brown and Jones make a series of predictive bets, they will not always make the same bet. They may *often* disagree on what, according to

16. "Different persons growing up in the same language are like different bushes trimmed and trained to take the shape of identical elephants. The anatomical details of twigs and branches will fulfill the elephantine form differently from bush to bush, but the overall outward results are the same" (Quine 1960, 8).

their chosen pattern, will happen next. To take a dramatic case, Brown may predict that Ella will decide to kill herself; Jones may disagree. This is not a trivial disagreement of prediction, and in principle this momentous difference may emerge in spite of the overall consonance of the two interpretations.

Suppose, then, that Brown and Jones make a series of predictions of Ella's behavior, based on their rival interpretations. Consider the different categories that compose their track records. First, there are the occasions where they agree and are right. Both systems look good from the vantage point of these successes. Second, there are the occasions where they agree and are wrong. Both chalk it up to noise, take their budgeted loss, and move on to the next case. But there will also be the occasions where they disagree, where their systems make different predictions, and in these cases sometimes (but not always) one will win and the other lose. (In the real world, predictions are not always from among binary alternatives, so in many cases they will disagree and both be wrong.) When one wins and the other loses, it will look to the myopic observer as if one "theory" has scored a serious point against the other, but when one recognizes the possibility that both may chalk up such victories, and that there may be no pattern in the victories which permits either one to improve his theory by making adjustments, one sees that local triumphs may be insufficient to provide any ground in reality for declaring one account a closer approximation of the truth.

Now, some might think this situation is *always* unstable; eventually one interpretation is bound to ramify better to new cases, or be deducible from some larger scheme covering other data, etc. That might be true in many cases, but—and this, I think, is the central point of Quine's indeterminacy thesis—it need not be true in all. If the strategy of intentional-stance description is, as Quine says, a "dramatic idiom" in which there is ineliminable use of idealization, and if Fodor's industrial-strength Realism is thus not the correct explanation of the reliable "visibility" of the pattern, such radical indeterminacy is a genuine and stable possibility.

This indeterminacy will be most striking in such cases as the imagined disagreement over Ella's suicidal mindset. If Ella does kill herself, is Brown shown to have clearly had the better intentional interpretation? Not necessarily. When Jones chalks up his scheme's failure in this instance to a bit of noise, this is no more ad hoc or unprincipled than the occasions when Brown was wrong about whether Ella would order the steak not the lobster, and chalked those misses up to noise. This is not at all to say that an interpretation can never be shown to be just wrong; there is plenty of leverage within the principles of intentional interpretation to refute particular hypotheses—for instance, by forcing their defense down the path of Pickwickian explosion ("You see, she didn't believe the gun was loaded because she thought that those bullet-shaped things were chocolates wrapped in foil, which was just a fantasy that occurred to her because . . ."). It *is* to say that there could be two interpretation schemes that were reliable and compact predictors over the long run, but that nevertheless disagreed on crucial cases.

It might seem that in a case as momentous as Ella's intention to kill herself, a closer examination of the details just prior to the fatal moment (if not at an earlier stage) would have to provide additional support for Brown's interpretation at the expense of Jones's interpretation. After all, there would be at least a few seconds—or a few hundred milliseconds—during which Ella's decision to pull the trigger got implemented, and during that brief period, at least, the evidence would swing sharply in favor of Brown's interpretation. That is no doubt true, and it is *perhaps* true that had one gone into enough detail earlier, all this last-second detail could have been predicted—but to have gone into those details earlier would have been to drop down from the intentional stance to the design or physical stances. From the intentional stance, these determining considerations would have been invisible to both Brown and Jones, who were both prepared to smear over such details as noise in the interests of more practical prediction. Both interpreters concede that they will make false predictions, and moreover, that when they make false predictions there are apt to be harbingers of misprediction in the moments during which thedénouement unfolds. Such a brief swing does not constitute refutation of the interpretation, any more than the upcoming misprediction of behavior does.

How, then, does this make me less of a realist than Davidson? I see that there could be two different systems of belief attribution to an individual which differed *substantially* in what they attributed—even in yielding substantially different predictions of the individual's future behavior—and yet where no deeper fact of the matter could establish that one was a description of the individual's *real* beliefs and the other not. In other words, there could be two different, but equally real, patterns discernible in the noisy world. The rival theorists would not even agree on which parts of the world were pattern and which were noise, and yet nothing deeper would settle the issue.[17] The choice of a pattern would indeed be up to the observer, a matter to be decided on idiosyncratic pragmatic grounds. I myself do not see any feature of Davidson's

17. Cf. "The Abilities of Men and Machines," in Brainstorms (1978), where I discuss two people who agree exactly on the future behavior of some artifact but impose different Turing-machine interpretations of it. On both interpretations, the machine occasionally "makes errors" but the two interpreters disagree about which cases are the errors. (They disagree about which features of the object's behavior count as signal and which as noise.) Which Turing machine is it really? This question has no answer.

position that would be a serious obstacle to his shifting analogies and agreeing with me. But then he would want to grant that indeterminacy is not such a trivial matter after all.[18]

What then is Rorty's view on these issues? Rorty wants to deny that any brand of "realism" could *explain* the (apparent?) success of the intentional stance. But since we have already joined Fine and set aside the "metaphysical" problem of realism, Rorty's reminding us of this only postpones the issue. Even someone who has transcended the scheme/content distinction and has seen the futility of correspondence theories of truth must accept the fact that *within* the natural ontological attitude we sometimes explain success by correspondence: One does better navigating off the coast of Maine when one uses an up-to-date nautical chart than one does when one uses a road map of Kansas. Why? Because the former accurately represents the hazards, markers, depths, and coastlines of the Maine coast, and the latter does not. Now why does one do better navigating the shoals of interpersonal relations using folk psychology than using astrology? Rorty might hold that the predictive "success" we folk-psychology players relish is itself an artifact, a mutual agreement engendered by the egging-on or consensual support we who play this game provide each other. He would grant that the game has no rivals in popularity, due—in the opinion of the players—to the power it gives them to understand and anticipate the animate world. But he would refuse to endorse this opinion. How, then, would he distinguish this popularity from the popularity among a smaller coterie of astrology?[19] It is undeniable that astrology provides its adherents with a highly articulated system of patterns that they *think* they see in the events of the world. The difference, however, is that no one has ever been able to get rich by betting on the patterns, but only by selling the patterns to others.

Rorty would have to claim that this is not a significant difference; the rest of us, however, find abundant evidence that our allegiance to folk psychology as a predictive tool can be defended in coldly objective terms. We agree that there is a real pattern being described by the terms of folk psychology. What divides the rest of us is the nature of the pattern, and the ontological implications of that nature.

Let us finally consider Churchland's eliminative materialism from this vantage point. As already pointed out, he is second to none in his appreciation of the power, to date, of the intentional stance as a strategy of prediction. Why does he think that it is nevertheless doomed to the trash heap? Because he

18. Andrej Zabludowski seems to me to have overlooked this version of indeterminacy in "On Quine's Indeterminacy Doctrine" (1989).

19. Cf. my comparison of "the astrological stance" to the intentional stance (Dennett 1989, 16).

anticipates that neuroscience will eventually—perhaps even soon—discover a pattern that is so clearly superior to the noisy pattern of folk psychology that everyone will readily abandon the former for the latter (except, perhaps, in the rough-and-tumble of daily life). This might happen, I suppose. But Churchland here is only playing a hunch, a hunch that should not be seen to gain plausibility from reflections on the irresistible forward march of science. For it is not enough for Churchland to suppose that, in principle, neuroscientific levels of description will explain more of the variance, predict more of the "noise" that bedevils higher levels. This is, of course, bound to be true in the limit—if we descend all the way to the neurophysiological "bit map." But as we have seen, the trade-off between ease of use and immunity from error for such a cumbersome system may make it profoundly unattractive.[20] If the "pattern" is scarcely an improvement over the bit map, talk of eliminative materialism will fall on deaf ears—just as it does when radical eliminativists urge us to abandon our ontological commitments to tables and chairs. A truly general-purpose, robust system of pattern description more valuable than the intentional stance is not an impossibility, but anyone who wants to bet on it might care to talk to me about the odds they will take.

What does all this show? Not that Fodor's industrial-strength Realism must be false, and not that Churchland's eliminative materialism must be false, but just that both views are gratuitously strong forms of materialism—presumptive theses way out in front of the empirical support they require. Rorty's view errs in the opposite direction, ignoring the impressive empirical track record that distinguishes the intentional stance from the astrological stance. Davidson's intermediate position, like mine, ties reality to the brute existence of pattern, but Davidson has overlooked the possibility of two or more *conflicting* patterns being superimposed on the same data—a more radical indeterminacy of translation than he had supposed possible. Now, once again, is the view I am defending here a sort of instrumentalism or a sort of realism? I think that the view itself is clearer than either of the labels, so I shall leave that question to anyone who stills find illumination in them.

2.5 Acknowledgments

Thanks to Kathleen Akins, Akeel Bilgrami, Donald Davidson, Barbara Hannan, Douglas Hofstadter, Norton Nelkin, W.V.O. Quine, Richard Rorty, George

20. As I have put it, physical-stance predictions trump design-stance predictions, which trump intentional-stance predictions—but one pays for the power with a loss of portability and a (usually unbearable) computational cost.

Smith, Peter Suber, Stephen White, and the MIT/Tufts philosophy of psychology discussion group for the discussions that provoked and shaped this paper.

References

Anscombe, G.E.M. 1957. *Intention.* Blackwell.

Chaitin, G. 1975. "Randomness and Mathematical Proof." *Scientific American* 232 (May): 47–52.

Churchland, P. 1979. *Scientific Realism and the Plasticity of Mind.* Cambridge University Press.

Davidson, D. 1989. "What Is Present to the Mind?" Presented at the Sociedad Filosófica Ibero Americana.

Davidson, D. 1991. "What Is Present to the Mind?" *Philosophical Issues* 1:197–213. https://doi.org/10.2307/1522929.

Dennett, D. C. 1969. *Content and Consciousness.* Routledge & Kegan Paul.

Dennett, D. C. 1978. *Brainstorms.* Bradford.

Dennett, D. C. 1981. "Three Kinds of Intentional Psychology." In *Reduction, Time and Reality,* edited by R. Healey, 37–62. Cambridge.

Dennett, D. C. 1982. "Beyond Belief." In *Thought and Object,* edited by A. Woodfield, 1 96. Oxford University Press.

Dennett, D. C. 1984. *Elbow Room.* 108–110. MIT Press.

Dennett, D. C. 1989. *The Intentional Stance.* MIT Press.

Dretske, F. 1988. "The Stance Stance." *Behavioral and Brain Sciences* 11: 511–512.

Fine, A. 1986. *The Shaky Game.* University of Chicago Press.

Fodor, J. A. 1987. *Psychosemantics.* MIT Press.

Gardner, M. 1970. "The Fantastic Combinations of John Conway's New Solitaire Game 'Life'." *Scientific American* 223 (October): 120–123.

Gardner, M. 1971. "On Cellular Automata, Self-Reproduction, The Garden of Eden and the Game of 'Life'." *Scientific American* 224 (February): 112–117.

Groot, A. D. de. 1965. *Thought and Choice in Chess.* Mouton.

Nelkin, N. 1994. "Patterns." *Mind and Language* 9 (1): 56 87.

Poundstone, W. 1985. *The Recursive Universe: Cosmic Complexity and the Limits of Scientific Knowledge.* Morrow.

Quine, W.V.O. 1960. *Word and Object.* MIT Press.

Sellars, W. 1954. "Some Reflections on Language Games." *Philosophy of Science* 21 (3): 204–228.

Sellars, W. 1963. *Science, Perception and Reality.* Routledge & Kegan Paul.

Sharpe, R. A. 1989. "Dennett's Journey Towards Panpsychism." *Inquiry* 32: 233–240.

Smith, P. 1988. "Wit and Chutzpah." Review of The Intentional Stance and Jerry A. Fodor's Psychosemantics, *Times Higher Education Supplement* (August): 22.

Wimsatt, W. 1980. "Randomness and Perceived Randomness in Evolutionary Biology." *Synthese* 43:287–329.

Zabludowski, A. 1989. "On Quine's Indeterminacy Doctrine." *Philosophical Review* XCVIII: 35–64.

3 Patterns All the Way Up: Prolegomena to a Future Naturalized Metaphysics

James Ladyman

Dennett's idea of real patterns was intended as a way of vindicating propositional attitude attribution within the scientific image, and it has been extended to understand the relationship between emergent entities and their components more generally. The notion of a real pattern has been applied to understand the ontology of various sciences, and it is central to the project of naturalized metaphysics for science as a whole undertaken in Ladyman and Ross (2007) (hereafter "ETMG"). Dennett's account is developed by David Wallace (2003) as part of a functionalist conception of emergent ontology applied to Everettian worlds. In that context, only higher-order entities are understood in terms of real patterns, whereas the wavefunction is understood in categorical rather than functional terms. On the other hand, Ladyman and Ross eschew the idea of a fundamental level and advocate a real patterns account of ontology across the board (including also merely notional or representational real patterns). Most other applications of the idea of real patterns in philosophy of science involve emergence from something other than fundamental physics. Vanessa Seifert analyzes the chemical bond as a real pattern in the charge density of valence electrons (2023). Daniel Burnston (2025) argues that neuroscientific models of decision-making are based on real patterns in the activation profile of the brain. Margarida Hermida (chapter 9) considers the tree of life as a real pattern in genealogical relationships, Ryan Nefdt (2023) argues that languages are real patterns in the behavior of some biological systems, and Wallace (chapter 8) discusses real patterns in physics that are not directly related to fundamental physics.

Dennett did not envisage the uses to which his work has been put in metaphysics, about which he was rightly skeptical, but at least those applying his ideas have shared his naturalism. He focused on the philosophy of cognitive science, which he thought about in Darwinian terms, and so he addressed the primary question of naturalized metaphysics, namely: How does science as a whole hang together? A central secondary question is: How do the ontologies of the different sciences relate, and, in particular, what is the place of physics within science as a whole? Dennett's teacher Quine argued that to be is to be the value of a bound variable, developing the idea that the criterion of existence is being quantified over in our best overall scientific theory, which he imagined as rendered in a purely extensional language, eliminating quantification over the objects of the special sciences in favor of physical objects and sets of them (1948)—the scientific image as desert landscape. This position amounts to antirealism about almost all of the ontology of science, including the ontology of physics, most of which deals with nonfundamental entities. Furthermore (as discussed below), the most fundamental current physics that is well confirmed and accepted is agreed to be incomplete, making for a blank scientific image with "watch this space" written on it. On the other hand, it is not possible to apply Quine's view ecumenically to scientific theories other than fundamental physics, because the objects of the special sciences are not all in a single domain of quantification. For example, a solid object like an iron bar can be modeled in continuum mechanics, but also with the atomic-lattice model of particles and bonds in which matter is not continuous. More generally the energy and timescales of the entities can be incompatible, so, for example, there is no ordinary matter at very high temperatures only plasma, or they are at different levels of abstraction as with individual organisms and species. Science cannot be thought of as conjoined into a single theory with a single domain.[1]

Dennett's idea of real patterns can be taken as an alternative to Quine's criterion of ontological commitment: To be is to be a real pattern, where a real pattern is something that indispensibly figures in projectible generalizations that predict and explain the behavior of the world. The sciences (and common sense) posit such objects, properties, relations, and processes that allow the formulation of such projections. Dennett's real patterns are computational. The compression of data and the reduction of information processing is made possible by a high-level description of a system that could, in principle, be described at a fine-grained level, but at a much greater computational cost. In Conway's Game of Life—Dennett's central example—the emergent objects—gliders, eaters—are mereologically composed of a small number of illuminated cells, and successive instances are composed of different cells, and successive instances a few steps apart have no cells in common. They persist only structurally.

1. Hence some naturalistic philosophers such as John Dupre (1995) embrace pluralism and the idea of multiple incompatible scientific images. On the other hand, below it is argued that the unity of science and the special place of physics within it must be acknowledged.

The idea of real patterns exemplified by the Game of Life has several important positive heuristics for naturalized metaphysics.[2] First, the idea of emergent patterns is generally applicable to how science represents the world. The days in which it was widely supposed that scientific theories could be rendered in first-order logic are long gone. On the alternative model-theoretic or semantic view of theories, scientific representation is to be understood as more analogous to a map than a set of sentences in a formal language, and mathematical representation is considered to be ineliminable. Scientific theories are very diverse, but all their ontologies are based on concepts that amalgamate mathematical and concrete concepts to describe emergent structure. The decision trees of evolutionary game theory are representations of a structure that only comes into being when there are populations of individuals whose choices and preferences approximate those of a particular game. Strategies and equilibria among them are aptly thought of as emergent real patterns. In chemistry the example of the chemical bond noted above can be augmented by the periodic table, which can be represented pictorially in different ways, and has no exact form since a few of the elements can be put in more than one place depending on which aspects of their chemistry are to the fore (Scerri 2019). The periodic occurrence of chemical properties expressed in the octet rule that preceded the periodic table is very aptly described as a real pattern. Such examples of emergent structure abound across the sciences. The wide applicability of the idea of a real pattern is in part due to its vagueness, but in particular contexts the idea can be made more precise in all the different ways that mathematical science has to describe patterns.

The second positive heuristic is that, in general, scientific representations often simplify the description of the phenomena. For example, the complexity of the positions at different times of the planets on the celestial sphere is reduced to Kepler's three laws of motion. Science is replete with relatively simple laws and principles. Of course, there is nothing simple about simplicity as much discussed in the philosophy of science, and in metaphysics, but nonetheless, in science, models and theories do often involve a compressed description of the data one way or another. Thirdly, the idea of real patterns implicitly involves scale-relativity, because the ontology of gliders and eaters is only applicable when more than a few cells over a few iterations is considered (different senses of scale-relativity and their significance for ontology

2. The terminology of positive, negative and, neutral heuristics roughly corresponds to that of positive, negative, and neutral analogies due to Keynes (1921) and Hesse (1966). ETMG argue that the task of naturalized metaphysics is to unify physics and the special sciences. It constructs a positive metaphysics based on ontic structural realism in the philosophy of physics, and rainforest realism about the special sciences using the idea of real patterns.

are discussed below). The analogue of the bit-map in the Game of Life is a description in terms of more basic ontology, and the gliders and eaters are analogous to higher-level ontology. For example, consider the ideal gas laws. Their ontology is that of gases somehow confined to a volume with their properties of pressure and temperature. The laws track real patterns in the behavior of gases regardless of how they are composed, hence they are multiply realizable.[3] The bitmap is the particles, and the gliders and eaters are the emergent properties of a gas. One real pattern is represented by Boyle's law, which states that the pressure is inversely proportional to the volume at fixed temperature. However, note that the law only fits the data approximately, and only when the volume is changed slowly. In general, the ontology of the gas laws is also scale-dependent in the sense that if there are only a few particles in a vast volume, there is nothing like an ideal gas, but there are still the particles and the laws they obey. The kinetic theory of gases offers a description of the behavior of the gas when the ideal gas laws break down. Furthermore, particle collisions are described by quantum theory as plane waves scattering off each other, and the particles in question, being atoms or molecules, are themselves composite entities.

In general, basing the idea of real patterns on emergence in the Game of Life has several negative heuristics for naturalized metaphysics as well.[4] Perhaps most importantly, the real mechanism that makes the system work is hidden in the implementation of the computation, the particular form of which is irrelevant. In the computational world there is a privileged fundamental level of description that is the bitmap, and then the emergent level of eaters and gliders. There is a finite number of cells, each of which is either on or off, and macrostates are completely determined by and reducible to microstates. Whether or not the world is fundamentally discrete and finite, the way in which the emergent entities are composed of the cells is not analogous to how parts compose wholes in science, insofar as the cells do not genuinely interact and their interactions do not compose the emergent entities which are themselves not genuinely causal. The gilders and eaters are analogous to life in only persisting structurally, but not otherwise, because real biological systems have to

3. There are two kinds of universality to the behavior of gases. First, each equilibrium macrostate is associated with numerous microstates, and secondly the same laws are obeyed by gases composed of different molecules or atoms, for example, hydrogen and helium, or a mixture like atmospheric air.

4. ETMG chapter 4 points out that the Game of Life is misleading as an exemplar of real patterns in some crucial respects. While the account of real patterns in Dennett, Ross (2000), and ETMG is articulated in information-theoretic terms, it can also be developed in terms of the dynamics of phase spaces, as Jenann Ismael suggests, and Ladyman and Ross (2013) and Alexseev et al. (chapter 10) argue that another way to understand it is in terms of statistical structures.

exploit physical and chemical mechanisms to survive. It is a one-dimensional model of reality in the sense that there is no independent way of identifying the real patterns other than in that particular kind of data. In real science, objects triangulate data from different sources. Moreover, in real science, there are cross-cutting classifications, and multiple domains and scales. For example, temperature manifests itself in numerous other contexts, unlike eaters and gliders.

There are many such lessons to be learned from current science. Some of the following propositions are more or less implicit in prior work by Ladyman and Ross and others, and many of them might seem obvious, but they are worth making explicit and compiling so that naturalized metaphysics can triangulate with them.

(1) The progress of science has involved both the expansion of ontology qualitatively and quantitatively in various domains and at various scales, and the integration of the models and theories of those domains and scales.

Many disputes in the history of science were resolved by expansions of ontology and scale. First Copernicanism required the universe to be much bigger than expected, and then following the "great debate" in the 1920s the existence of galaxies ("island universes") other than the Milky Way was accepted (and it now seems likely that there is much more dark matter than ordinary matter). In chemistry the number of elements is over ninety, not four or five, and the number of possible compounds is truly vast. In biology the discovery of microbes was followed by the realization that not only do they constitute most of the biome by mass, but there is an entirely new taxon of Archaea. Overall, there are vastly more entities and types of entities in current science than in the past, all based on the same core ontology of the elements of the periodic table, plus the fields and forces of physics, and the standard physical quantities and units. As well as more individual things and types of things, there are also many more properties, processes, and relations. Properties such as fitness, processes such as adaptation, and relations such as being conspecific, are peculiar to biology, and quite different from those in chemistry such as solubility, oxidation, and isomerism. On the other hand, all these entities are understood in an interconnected way, so that the chemical entities mentioned are all relevant to biological systems. Accordingly, there are many boundary disciplines such as biochemistry and geophysics. Even critics of reductionism, such as Nancy Cartwright (1999) and Sandra Mitchell (2003), emphasize that science is integrated. Astronomy, biology, chemistry, geology, and medicine were once separate subjects, but now they are deeply interconnected.

(2) Ontological commitment in science depends on triangulation across different modalities of measurement, and across different scales and domains.

There are two related ideas of triangulation. The first is the epistemological idea that scientific theories should be supported by multiple independent lines of evidence, which has been central to thinking about the scientific method since at least the writings of Robert Boyle. The second is the ontological idea that a real object manifests itself in more than one way, which is true of everyday objects and also the objects of science (Hacking 1983). In the best epistemic case, the multiple lines of evidence include multiple kinds of evidence in different domains. For example, many different terrestrial and extraterrestrial data confirmed Newton's laws of motion and gravitational force law, and this remains a prime example of what Whewell called "the consilience of induction" (Laudan 1971). There can be multiple independent kinds of evidence for an object that manifests itself in more than one way. The atomic theory of matter was universally adopted only when it triangulated in chemistry and physics in many domains and at many different scales (Chalmers 2009), multiple independent precise determinations of Avagadro's number being decisive for some holdouts such as Poincaré: "The brilliant determinations of the number of atoms computed by Mr. Perrin have completed the triumph of atomism. What makes it all the more convincing are the multiple correspondences between results obtained by totally different processes" (1963 [1913], 90). Similarly, tectonic plates were originally proposed to explain the large length and timescale of geological and paleontological data, but they were eventually accepted in the light of data about the current structure of magnetic field lines on the bottom of the oceans (Frankel 2009). Cosmic rays were identified as charged particles, not photons, because their intensity was found to be affected by the magnetic field of the Earth (Hartog 1949). The ontological commitments of the sciences in general, such as cells, galaxies, hurricanes, molecules, people, and social structures are epistemically on a par with those of physical science insofar as they triangulate across different modalities of measurement, and different scales and domains too. Triangulation in this sense is very similar to Ruth Millikan's idea of "clusters" which she understands as nonaccidental (2017, 13).

(3) The diversity of the ontology of science is paralleled by the diversity of the kinds of causation, equilibria, laws, mechanisms, powers, probabilities, principles, and symmetries in science.

Aristotle's four kinds of explanation have given way to a multitude. The varieties of scientific explanations now include the explanation of the adaptation of the eye by natural selection, the explanation of the geopolitics of the Cold War in terms of the Nash equilibrium of mutually assured destruction, the explanation of the abundance of iron in the universe by the stability of its nucleus, the explanation of the falling of a cat by the law of conservation of angular momentum, the explanation of the firing of nerve cells in terms of action potentials, the explanation of acidity in terms of valence electrons, the explanation of the stability of matter by the Pauli exclusion principle, and the probabilistic explanation of dosimetric data provided by nuclear physics. Causal explanations alone defy the many ingenious attempts at unitary analysis, so causation is plausibly of various kinds, both in respect of causal concepts and their targets in the world (Cartwright 2004, Godfrey-Smith 2010). All the other kinds of modality mentioned above are discussed in the metaphysics of science literature, and the idea that they can all be reduced to one kind is even more implausible. On the other hand, there is no naturalistic motivation for eliminating modality from science altogether (whether by understanding all explanations as subsuming particular facts under regularities, or as fictions), because all scientific ontological commitments are based on the identification of something in a local environment relative to which it has some kind of stable modal profile. Successful scientific representations are projectible, supporting explanation, prediction, and counterfactual inference, and there are no real things without projectibility, so an ecumenical attitude to the modalities of science as well as its ontology is appropriate. Indeed, the two go together since, on the real patterns understanding of emergence, objects, and their modal profiles are inseparable, just as, for example, electrons are inseparable from Pauli's exclusion principle and the laws of quantum electrodynamics. The blurring of the distinction between objects and modal structure is well exemplified by stars, because they are understood as individual objects by being understood as processes, and by galaxies, which maintain a steady state over billions of years by stars being born and dying constantly within them. Similarly, macroscopic objects are understood as relatively stable and enduring real patterns that emerge from the myriad interactions of their component atoms and electrons.

(4) The concept of natural modal structure is an abstraction that includes all of the above. In formal mode this modal structure is absolute, but in material mode it is domain- and scale-dependent.

Modal structure is the genus to which the different species of causal structure, nomological structure, and so on belong, and all may also be thought of as kinds of real pattern. They are not to be thought of as mere regularities because the best scientific explanations have in common that they are associated with reliable predictions and counterfactuals. For example, Newton's law of gravitation made numerous predictions and tells us that the Earth's orbit would be less stable without the Moon. A general form of realism about science is expressed by the commitment to the reality of the modal structure tracked by our best scientific theories.[5] Novel predictive success, such as the detection of the Higgs Boson and gravitational waves, has long been regarded as the strongest possible form of confirmation, and yet none of the most predictively successful theories to date are universally true in all domains at all scales. When we attribute modal structure with a scientific model (in formal mode), we represent it as obtaining indefinitely in space and time, ignoring many aspects of systems and their environments (in material mode). Consider the model of the harmonic motion of a pendulum. Real pendula only ever behave like that over some time period, in respect of some of their degrees of freedom, and only approximately. Furthermore, any causal or lawlike structure is attributable to a real system only insofar as it and its local environment are not subject to a significant perturbation. Every attribution of modal structure to systems on the earth is modulo the nonoccurrence of an imminent asteroid strike, and all other modal structures break down at very high temperatures and in the very early universe, as well as in exotic astrophysical conditions such as within neutron stars. No real system obeys time-translation invariance if the transformation is big enough because it has a finite past, not just because the materials from which it is made are produced in complex processes but because elements more complex than hydrogen and helium only exist in abundance after more than one generation of stars. Correspondingly, every physical system on Earth has a finite future if only because the Sun will eventually die.[6]

(5) There is modal structure to the way that domains and scales relate.

5. Ladyman and Ross argue on the basis of consilience that natural modal structure is represented by science. Indeed, they define real patterns modally. Probability and statistics, and computation and information are all modal and all ubiquitous in science. Realism about modal structure is a core commitment of ontic structural realism (Ladyman 2023).

6. Physics studies particular objects as well as generalizing, and so physics is also partly an historical science. This has always been true of parts of solar system physics and geophysics, but now the actual history of the whole universe is modeled in general relativistic cosmology.

There are connections between domains and scales that are not accidental and arbitrary but projectible and systematic. Correspondingly, when the modal structure represented by a theory breaks down, this can often be described and modeled in terms of another theory. For example, the laws of ray optics are approximately true for the behavior of light when its wavelength is small relative to the things with which it is interacting. The relations between the domains and scales of ray optics and wave optics are structured by the nature of light, and the more fundamental laws hold when the less fundamental ones do not—for example, when there is diffraction, or when the light is considered in conditions in which it behaves like photons and not like a classical wave. Sometimes the connections take the form of law-like relations, and, in general, in physics theories about different domains are often related by mathematical limits, as with special relativistic and classical mechanics. There is also modal structure to the relations between domains and scales in the form of affordances and constraints—for example, the modal structure of the relations between physics and biology, such as the biological affordances of the large surface area to volume ratio of a wing, and the constraints of a large mass on the shape of wings in birds compared to insects (Hermida and Ladyman forthcoming). The affordances and constraints governing reactions among chemical substances reflect the modal structure of the electromagnetic force between charged particles, the electronic shell structure of atoms, and the spatial structure of molecules. Science includes theories that link the levels, but in almost all cases the higher-level description is spatially and temporally coarse-grained, approximate, and aggregative with respect to the underlying description, so reduction based on type-type or token-token identity relations between higher-level and lower-level entities is not plausible. For example, the pressure of a gas is an average property of a volume of gas over some short time, and it moves pistons, which a set of instantaneous microstates cannot do, so the macroscopic property of pressure is not identical with any of the specific microstates associated with it in statistical mechanics nor with any property or set of them.[7]

(6) Sometimes the modal structure of the relationships between domains and scales is due to composition or part-whole relations, but these relations are in general diachronic, dynamical, and involve domain-specific interactions and relations among parts.

7. ETMG argues against ontological reduction understood as requiring either type-type or token-token identity. There are various accounts of reduction in current philosophy of physics that do not require either. Clearly naturalistic metaphysics must acknowledge the success of reductive projects in science of which there are many as noted above.

Richard Feynman (1963) said that the single most informative proposition in science is that everything is made of atoms. This is apt to mislead because, in current metaphysics, composition is usually understood in terms of mereology, which considers part-whole relations in general. It tells us that parthood is antisymmetric and transitive, and there are various interesting formal systems with further features. However, it does not include interactions and other relations, whereas scientific accounts of composition almost always involve them. In the case of the atomic-molecular theory of matter, the behavior of gases and liquids obviously depends on the dynamics of their parts, but even the seemingly fixed structure of solids is due to their electromagnetic interactions. Furthermore, ultimately the electrons and nuclei that compose them are dynamical entities, and so atoms are not simple mereologically or otherwise. Proteins only have their functional properties because of both the sequence of their component amino acids and their shape. The immune system has its functional properties due to its component entities and the signals they send to each other. Organisms compose an ecosystem because they interact, likewise for cells and multicellular organisms. This introduces various forms of scale-dependence. Most composite systems with emergent features have large numbers of parts that interact many times. Anderson's (1972) "more is different" in condensed matter physics refers not only to particles but also to their interactions. In complex systems in general, order and self-organization can emerge spontaneously from the repeated interactions among parts. There is often a very large separation of timescales associated with emergence. For example, the interactions of the atoms or molecules in a gas take place on a very short timescale compared to the timescales over which macroscopic properties are operationally meaningful. Despite rhetoric about emergent phenomena defying reductive explanation, complexity science involves breaking things into parts and studying their interactions to understand the behavior of the whole. Order and organization can arise even in simple nonliving systems when they are driven exogenously in some way, so that the effect of the environment on them cannot be ignored (Ladyman and Wiesner 2020).

(7) The modal structure of the relationships between domains and scales can be construed in terms of dependence, determination, or supervenience.

For example, the laws of stellar evolution depend on the laws of nuclear physics, the hardness of diamond is determined by its tetrahedral unstrained bonds, and the software states of a computer supervene on the states of those of its physical degrees of freedom that are implementing the computation. In discussing the relationships between different domains and scales in science terms such as "based on," "gives rise to," and "fundamental" are also used. While philosophers often seek to regiment such terminology purely linguistically, scientists do so with mathematics, models, and theories.

(8) Physics is the science that covers the widest range of scales and regions of space and time.

Biology only applies where and when there is life, and the social sciences only where there are collections of people interacting. Chemistry only applies at energies at which nuclei can capture electrons in their outermost shell (valence electrons), so there is no chemistry at very high temperatures. According to the standard model of cosmology, in the very early universe there was no matter other than hydrogen and helium gas and, before that, there were not even protons and neutrons. Later, when black holes have formed, there are regions of very high curvature that are also incompatible with the existence of ordinary matter. Electromagnetism applies to molecules and to pulsars, and from the shortest to the longest length scales, but it breaks down at the energy when it becomes unified with the weak interaction. Hence, although a lot of physics applies with very great generality, ultimately all of physical science is scale-dependent or domain-specific to some extent. Arguably the most generally applicable part of physics is thermodynamics, which is critical for chemistry, biology, and all of physical science.

(9) The modal structure of the relations between domains and scales reflects the fact that materialism has been successful as a research program in science.

The programs that created new physical ontology and processes exclusively to account for chemical, living, and cognitive systems have not had any scientific success, while programs based on understanding such systems in terms of more fundamental physical ontology and processes have been successful. The modal structures of the different sciences all stand in modal relations with the modal structures of physics directly, as in biophysics, and indirectly as in the relation of endocrinology to biochemistry to physics. Among the most successful reductive programs is that of chemistry to physics insofar as chemical properties can now be predicted by the application of electromagnetism and quantum mechanics to molecules. In biology the functional profile of proteins is based on the amino acids they contain and their shape, which determines how they interact with other molecules. Furthermore, physical features of systems, and physical measurements, are central to the rest of science. This is why ETMG took naturalized metaphysics to be demarcated by the principle of naturalistic closure (PNC):

> If a metaphysical claim is to be taken seriously at a time *t*, it should, if true, show how two or more *scientific hypotheses*, at least one of them *specific* and at least one of them both specific and drawn from fundamental physics taken seriously at *t*, jointly explain more than the sum of what is explained by the two hypotheses taken separately. (Ladyman and Ross 2007, 37)

However, this is problematic because it makes reference to fundamental physics. It might be modified to refer instead to "a more fundamental theory," where that will often be a theory from physical science.

(10) The physics that is fundamental with respect to the special sciences, and which unifies them, is not absolutely fundamental, and there is no single fundamental physics.

The elements of the periodic table and the four fundamental forces of nature are not all ultimately elementary. Furthermore, in the current scientific image there is not a single privileged level of description, because there is no single fundamental physics, rather there is the Standard Model and general relativity. Dark energy and dark matter, as well as quantum gravity, are largely mysterious.

(11) There is no single hierarchy of levels, nor are there sharp boundaries between domains and scales. It also should not be assumed that there is a single fundamental level.

Pictures of reduction and emergence divide the world into levels, and while there is structure at many scales, there is not all kinds of structure at all scales. Informal talk of levels of ontology should not be understood as representing distinct global levels of existence. There is no single hierarchy because there are various different kinds of relationships between lower-level and higher-level entities. Sometimes different levels correspond to distinct but related properties—for example, the conductivity of a metal, and the dissociability of its electrons. Levels may also be associated with different length scales, but need not be—for example, the different levels of description in micro and macro-economics. Sometimes different levels correspond to a part-whole relation as with the levels of cells and organisms, but of course some organisms are unicellular. It may also be useful to roughly divide the world into the living and the nonliving. In specific models and theories there may be also be a distinction between levels, as for example, the levels of microstates and macrostates in statistical mechanics. The idealization of separate levels often depends on the effective separation of the timescales of the relevant dynamics—for example, in the separation of the electronic and nuclear dynamics due to their vastly different masses and the separation of the dynamics of the interactions of the cells in a muscle fiber from the dynamics of a running animal.

However, the idea of a single hierarchy starting with subatomic physics and going up through chemistry to biology and psychology does not correspond to anything in scientific practice. Theories and models in science often cut across scales to some extent, and as noted in (2) above, ontological commitment in science usually requires that entities in question triangulate different evidence from different domains and at different scales. Physics ranges from the scale of quarks and leptons, to that of superclusters of galaxies, but high-energy physics is applied to understand both the minutiae of collisions in particle accelerators and to the life cycle of stars. In physics the low-energy is often the emergent, and the higher-energy the more fundamental, and this also often coincides with bigger and smaller, respectively, and with the whole-part relation. However, they come apart in some contexts, including when it comes to the most fundamental physics, as higher-energy corresponds to smaller length scale in particle physics but not in gravitational physics. In quantum mechanics, more fundamental does not mean smaller because of entanglement. The levels of ray optics, wave optics, electromagnetism, and quantum electrodynamics do not correspond to the levels of sub-atomic, atomic, and molecular. Molecules can be microscopic, mesoscopic, and macroscopic. Even within physical science there is no single hierarchy of fundamentality, and hence no determinate notion of the ultimate fundamental level.[8]

The vast ontology of biology covers scales from the microscopic to the planetary, including many functional entities that are not part of the ontology of chemistry and physics, but also including nonliving things such as amino acids and carbon dioxide. Iodine is part of the ontology of endocrinology, but so is the thyroid gland. The levels of functional organization in biology do not correspond to differences in length that are relevant to physics, and in biological systems, processes at very different scales are coupled in ways that depend on their histories. The boundary disciplines are necessary because of the multiple and vague interfaces between domains and scales as discussed.

8. ETMG argued that there is not, for all we know, a fundamental level of reality (though there might be), and that accordingly our metaphysics should not presuppose that there is. Many commentators took the claim that relational structure is fundamental to be a form of fundamentalism (for example, McKenzie 2017), but it was not so intended.

(12) All existing theories have an effective ontology of emergent entities. Effective ontology is scale-dependent.

Sokal and Bricmont (1997) defend what they call the renormalization group view of the world, and ETMG claims that, in general, ontology is scale relative in respect of both space and time. There are different kinds of scale, including lengthscale, timescale, numerosity of parts, and numerosity of interactions, and different domains associated with different degrees of freedom. The idea of scale-relativity and effective ontology is standardly discussed in the context of the quantum field theories of the standard model because they break down at very high energies. However, thinking of the objects of physics in a domain or scale-dependent way goes back much further. Arguably, the application of the law of the lever requires that the physical object being used is thought of as a rigid body, whereas in fact it is more or less obviously distorted to some extent by the application of a force. In other words, the law of the lever applies when the body used is effectively rigid when subjected to the forces being applied and not otherwise, and so the law is effective. Every object, property, or event is classified in a way that does not capture modal structure at all scales, where energy scales are also relevant and correlated with length and timescales in interesting ways. For example, over very long timescales, not even the most enduring and natural ordinary objects existed, because in the early universe there was no ordinary matter. Similarly, in the future, the Earth will be absorbed by the Sun when it becomes a red giant. Everything in chemistry depends on thermodynamic equilibria and free energy, both of which are dependent on pressure and temperature. The laws of chemical reactivity are therefore all scale-dependent.[9] More fundamentally, the elements and compounds with which chemistry is concerned do not exist at temperatures at which electrons have enough energy to escape being bound by nuclei. The most perfect vacuum ever made in a laboratory on Earth has a higher density of atoms and molecules than gases in astrophysics, so whether or not a region of space is thought of as occupied by a gas is scale-relative, and a matter of whether it is effectively described as such.

If the idea of effective ontology sounds too vague for proper science here is one of the greatest mathematical physicists of all time in his book on classical mechanics: "A body so small that, for the purposes of our investigation, the distances between its different parts may be neglected, is called a material

9. For other examples of scale-relativity in chemistry, see Hendry (2021) and Ladyman and Seifert (forthcoming).

particle" (Maxwell, 1888 [2010] 3). This means that a planet may be a particle for some purposes. Maxwell here refers to purposes, but the constraint on what is effective in what context comes from the world. Classical mechanics is completely useless for describing the behavior of electrons. Vagueness is a consequence of effective ontology because the usefulness of a model often breaks down by degrees. A pendulum subject to significant friction may not look very harmonic for very long. The number of tectonic plates depends on exactly how they are defined, as does the number of layers of the atmosphere and where exactly they start and stop. However, vagueness is not complete indeterminacy, and everybody agrees about the major tectonic plates and the major atmospheric layers, though seeking to define either to within a millimeter would display complete ignorance about the relevant domain and the characteristic scale of its features. Accepting the idea of effective ontology helps reconcile scientific realism with theory-change, and also to some extent with common sense realism insofar as many everyday objects, most obviously material objects like tables, have at least some of their properties as effective scientific entities in the scientific image.[10] Effective realism reconciles the scientific descriptions of different domains and scales. There are real patterns in the world that are only discoverable at the right scales of resolution and degrees of approximation. Not recognizing them is missing something about reality, and that is enough to say that the objects, properties, and processes described by the special sciences are real.[11]

(13) Downward causation understood as the intervening of the higher level in the lower level directly is not coherent, because there are no distinct levels, and because the higher-level entity acts via its parts.

Scientific images founded on a fundamental physical level seem to leave no room for causation in the rest of science. Naturalistic philosophers often embrace ideas of downward causation in order to vindicate the genuine causal power of entities in the special sciences, and hence the genuine reality of the emergent entities of the special sciences (emergence without causation being considered merely epistemic). However, the difference making causes that are sought in the special sciences are often at the higher-level because the difference in question is to an explanans that refers to higher-level facts. The

10. See Ladyman (2018), and Ladyman and Lorenzetti (forthcoming).

11. The term "special sciences" is usually used to refer to everything other than physics, but for the purposes of naturalized metaphysics, most of physics is a special science because it deals with effective emergent entities.

relevant difference making cause of the increase of the pressure of a gas is the increase in temperature even though any change in temperature is associated with a change in the microstate. In the special sciences one is usually interested in "universal" forms of behavior, where "universal" means independent of the detailed behavior of microphysical or other kinds of constitutive states. The identification of universality and the appropriate descriptive categories for tracking it is one of the principal tasks of the special sciences.

(14) In science it is sufficient, and usually necessary, to track only the statistical properties of low-level entities when tracking high-level entities. Special science ontology involves abstraction, aggregation, approximation, and coarse-graining.

The special sciences are possible insofar as the data in a given domain are, to some extent, algorithmically compressible. The models and theories of the special sciences track the emergent collective behavior of systems using fewer degrees of freedom than would be needed to describe the underlying state of the system (Wilson 2010). For example, in condensed matter physics, the renormalization group describes transformations that allow the number of degrees of freedom in the Hamiltonian of a system to be massively reduced while still recovering the critical behavior of the system. The reduction in the number of parameters needed effectively to describe systems is exactly what the theory of real patterns aims to capture. However, scientific real patterns are lossy in the sense that the compression of the information allows only approximate or probabilistic recovery of the behavior of the underlying system (unlike the real patterns in the Game of Life). It is possible to use much less information to predict the behavior of systems described in an approximate and probabilistic way than would be needed to describe everything about them. Emergence in physics is often described in terms of "coarse-graining" with respect to underlying degrees of freedom. The ideal gas laws use only three degrees of freedom to give a pretty good description of the behavior of systems that have of the order of 10^{23} degrees of freedom. Hence, changes in pressure or temperature are difference making causes with respect to the coarse-grained ontology. This is another reason why even token identity does not obtain. If one ignores emergent real patterns, one misses out on a real feature of the world.

(15) The idea of a description of the whole universe at all scales has nothing to do with any scientific theory that we could develop, test, and use.

Any actual scientific theory can only be about part of the universe in some respects at some scales, because it is used by agents that cannot include a complete description of themselves in the model (this is why the definition of real patterns in ETMG refers to perspectives). It is also necessary that the agents have information channels to the system and its environment, and internal states with which they can compute. The agents assume that their effects of the rest of the universe can be ignored, and that the rest of the universe can be separated into a system and its local environment. It is also often but not always assumed that the effect of the system on the environment can be ignored, as with the notion of a heat bath, which is considered so large compared to the system that any heat flow to or from the system to it does not change its temperature. It is also often but not always assumed that the effect of the environment on the system can be ignored, so that the system can be modeled as if it was the only thing there is.

References

Anderson, P. W. 1972. "More Is Different." *Science, New Series* 177 (4047): 393–396.

Burtson, D. 2025. "Nagelian Reduction and the Concept of 'Decision'." *Synthese* 205 (74).

Cartwright, N. 2004. "Causation: One Word, Many Things." *Philosophy of Science* 71 (5): 805–819.

Chalmers, A. 2009. *The Scientist's Atom and the Philosopher's Stone: How Science Succeeded and Philosophy Failed to Gain Knowledge of Atoms.* Springer.

Dupré, J. 1993. *The Disorder of Things: Metaphysical Foundations of the Disunity of Science.* Harvard University Press.

Frankel, H. 2009. "Plate Tectonics." In *The Cambridge History of Science,* edited by P. J. Bowler and J. V. Pickstone, 383–394. Cambridge University Press.

Godfrey-Smith, P. 2009. "Causal Pluralism." In *The Oxford Handbook of Causation,* edited by H. Beebee, C. Hitchcock, and P. Menzies. Oxford University Press.

Hacking, I. 1983. *Representing and Intervening.* Oxford University Press.

Hartog, H. den. 1949. "Clay's Work on Cosmic Rays." *Synthese* 8: 3–5.

Hendry, Robin Findlay. 2021. "Structure, Scale and Emergence." *Studies in History and Philosophy of Science Part A* 85: 44–53.

Hesse, M. B. 1963. *Models and Analogies in Science.* University of Notre Dame Press.

Keynes, J. M. 1921. *A Treatise on Probability.* Dover Publications.

Ladyman, J. 2017. "An Apology for Naturalized Metaphysics." In *Metaphysics and the Philosophy of Science,* edited by M. Slater and Z. Yudell. Oxford University Press.

Ladyman, J. 2018. "Scientific Realism Again." *Spontaneous Generations* 9 (1): 99–107.

Ladyman, J., and M. Hermida. Forthcoming. "Physical Explanation and the Autonomy of Biology." *Philosophy of Science.*

Ladyman, J., and L. Lorenzetti. Forthcoming. "Effective Ontic Structural Realism." *British Journal for the Philosophy of Science.*

Ladyman, J., and D. Ross. 2007. *Every Thing Must Go.* Oxford University Press.

Ladyman, J., and D. Ross. 2013. "The World in the Data." In *Scientific Metaphysics,* edited by H. Kincaid, J. Ladyman, and D. Ross, 108–151. Oxford University Press.

Ladyman, J., and V. Seifert. Forthcoming. "Scale Relativity in the Metaphysics of Chemistry."

Ladyman, J., and K. Wiesner. 2022. *What Is a Complex System?* Yale University Press.

Laudan, L. 1971. "William Whewell on the Consilience of Inductions." *The Monist* 55 (3): 368–391.

Maxwell, J. C. 1888. *Matter and Motion.* [2010]. Cambridge University Press.

McKenzie, K. 2017. "Ontic Structural Realism." *Philosophy Compass* 12 (4): e12399.

Millikan, R. G. 2017. *Beyond Concepts: Unicepts, Language, and Natural Information.* Oxford University Press.

Mitchell, S. D. 2003. *Biological Complexity and Integrative Pluralism.* Cambridge University Press.

Nefdt, R. M. 2023. *Language, Science, and Structure: A Journey into the Philosophy of Linguistics.* New York.

Poincaré, H. 1963. *Science and Mathematics: Last Essays.* [1913]. Dover Publications.

Quine, W.V.O. 1948. "On What There Is." *Review of Metaphysics* 2 (5): 21–38.

Scerri, E. R. 2007. *The Periodic Table, Its Story and Its Significance.* Oxford University Press.

Seifert, V. A. 2023. "The Chemical Bond Is a Real Pattern." *Philosophy of Science* 90 (2): 269–287.

Sokal, A. D., and J. Bricmont. 1998. *Intellectual Impostures: Postmodern Philosophers' Abuse of Science.* Edited by J. Bricmont. Profile Books.

Wallace, D. 2003. "Everett and Structure." *Studies in History and Philosophy of Science Part B: Studies in History and Philosophy of Modern Physics* 34 (1): 87–105.

4 Abstractions by Patterns

Steve Petersen

Daniel Dennett's "Real Patterns" paper (1991) seeks in part

> to convince philosophers that a mild and intermediate sort of realism is a positively attractive position, and not just the desperate dodge of ontological responsibility it has sometimes been taken to be. (29)

His paper specifically defends mental states as a candidate for "intermediate realism," but I think the trick generalizes: There are many cases where patterns of more basic things are *real enough*, and we should adopt similar intermediate realisms for them too. For example, philosophers get vexed over whether composite objects "really" exist, somehow "over and above" their parts. Some say no, there are not really any objects made out of other objects; others say that any collection of objects makes a new object in good standing. Both positions seem prima facie absurd, and to resolve this dilemma I've argued in the past that *patterns* of information in the more basic things can distinguish some collections, but not others, as deserving of being called a new thing—one that is real enough.[1]

I call this the "patternist" approach, and I suspect we can profitably apply it to a cluster of concepts even more notorious for ontological dispute: Concepts like *properties*, *universals*, and *natural kinds*. The debate over whether such things are "real" or not has an ancient pedigree, which many trace back to a fundamental dispute between Plato and Aristotle. The long-standing disagreement suggests, once again, that Dennett's intermediate realism via patterns might be in order.

The notion of "pattern" at play has always been about *information compression*. In my telling, a pattern in some data is any regularity that allows for

1. Petersen (2019). For more on the question of the existence of ordinary objects (made out of smaller things like atoms), see Korman (2011). The central question—the "Special Composition Question" of when a plurality of things makes a new thing—is laid out in van Inwagen (1990). Van Inwagen's is a rival "moderate" account that, like mine, tries to walk the line between "no collection makes a new thing" and "all collections make a new thing." (I think my proposal subsumes his.) Bird (2023) is a more recent, similar proposal that largely endorses mine.

lossless compression of those data. I propose, in summary, that some sets of objects are more natural than others, in virtue of compressibility facts. Since goodness of set compressibility can differ in kind as well as degree, we can delineate levels of robustly "natural" sets deserving of names like "universals" or "natural kinds" from more heterogeneous "sets" or "properties." In this chapter, I will illustrate the approach with a focus on *abstractions*.

True Dennett fans may, at first glance, find this project intolerably metaphysical. Scholastic quibbles about the nature of universals seems against the spirit of Dennett's American pragmatism, which is partly characterized by putting philosophy on a continuum with the sciences. I am optimistic, however, that my proposals can shed light on controversies of interest to scientists. After giving my patternist account of abstraction, this chapter then shows how it can illuminate more straightforwardly scientific notions like explanatory inference and emergence.

My approach will thus be revisionary; it is not straightforward conceptual analysis, and I will not pretend to capture every intuition that anyone has about abstraction. Ultimately, I am making *proposals* for understanding abstraction and nearby concepts—proposals that should be judged by their potential use in philosophy and science. If they are near enough to the pre-revisionary forms, they might usefully supplant those terms, at least in the right contexts; if not, the proposals could still stand on their own with neologisms in their place, to make their differences explicit from the older, more sacrosanct notions. In other words, if you prefer, you can imagine that I'm stipulating definitions for "patternist-abstractions," "patternist-emergence," and so on, and then assess those on their merits.

4.1 Philosophical and Formal Preliminaries

We need first to lay out the general framework of patternism, including the formal definition of a pattern. I start, though, by motivating the general algorithmic method it employs.

4.1.1 Information Theory and Philosophy

The formal definition of a pattern—gestured at briefly in Dennett's original paper—comes from *algorithmic information theory* (AIT).[2] AIT studies the complexity of and information contained in binary strings, which are just finite but arbitrarily long concatenations of zeros and ones, each of which is called

2. Dennett's brief reference is in footnote 10 (original publication) or footnote 4 (this volume). The field is also called "Kolmogorov complexity theory," after one of the main founders, Andrey Kolmogorov; Dennett refers to it via a different founder, Gregory Chaitin. (The third main founder is Ray Solomonoff.) The classic text is Li and Vitányi (2008).

a "bit." (You can also think of the bits as "on"s and "off"s, or "yes"s and "no"s.) Binary strings may seem like uninteresting objects of study, until we remember that all the precious files we keep on our computers and phones are really just such binary strings. Indeed it seems just about anything can be represented to arbitrary precision with enough information, which is to say it can be "digitized" with a long-enough binary string. For example, if you think Nick Bostrom's "simulation hypothesis" is at least coherent—if you think it's *possible* our existence is realized in a digital computer running in a reality one level deeper than ours—then plausibly you already think not just that all the ordinary objects and people with which we're familiar could be *represented* arbitrarily well by digital strings, but rather that those objects and people might themselves just *be*, in an important sense, digital strings.[3]

Fellow Dennettians Don Ross and James Ladyman (2007) tentatively endorse what they call "a deflationary information-theoretic foundationalism," quoting Anton Zeilinger:

> It is impossible to distinguish operationally in any way reality and information . . . the notion of the two being distinct should be abandoned.[4]

I too tentatively endorse this claim: I think science can successfully proceed as though reality and information were the same. This principle, if correct, allows us to apply powerful algorithmic tools to the study of reality.

The binary strings at issue could be encodings of a particular subject's total sensory inputs, or of all intersubjective scientific "observations" in a community—or even of the totality of ground-level, objective facts, as in David Lewis's "Humean mosaic" (1986). What one considers the proper source of data will thus depend partly on one's empiricist or realist taste. I will be neutral about what the binary strings might represent (or *be*), treating them simply as "data."

4.1.2 Patterns Defined

The intuition behind a *pattern* in the sense at hand is a kind of regularity in the data—which is to say, a sufficient amount of mutual interdependence among the data that allows us to infer, better than chance, some of the data based on the rest. For example, when humans see a string like `011011011011011`, they have a pretty good guess of what bits come next. If some of the data

3. See Bostrom (2003) for the simulation hypothesis (and the case that we probably *are* in such a simulation), and see Chalmers (2005) and Chalmers (2022) for arguments that "real" things could be fundamentally digital in this way (rather than fundamentally eleven-dimensional strings, for example, as some physicists propose).

4. Zeilinger (2004), 219, which Ladyman and Ross quote on p. 189.

are enough to make a good guess about the remainder, then this allows for lossless compression of the data, which is at the heart of the formal definition. To say that some data (a string of bits) is losslessly compressible by a computer language is to say that I could send you a program in that shared language that is shorter than the target string, and that would recreate the target data perfectly. (A "program" of course is just a *different* string of bits.) As the canonical example goes, instead of sending you a data file made of one trillion `1`s—which would take up 125 gigabytes, and require about an hour and a half to transmit on a decently quick internet connection of 2024—I can send you the same file using a much shorter, self-contained program like `for i from 1 to m: print 1`, plus the input parameter of `m=1,000,000,000,000`, which together would reproduce the original file on your end exactly.[5] In general, a lossless compression of x involves a *program* p, and an *input* to that program n. In the example above there was one input, but encoding tricks can let us mash a bunch of inputs into one ordered n-tuple representation of them. So for example the program could have had instead two variables, `m` and `j`: `for i from 1 to m: print j`, and then have as input a digital string encoding the pair <`m=1,000,000,000,000, j=1`> (which the program is designed to parse).

That is the informal idea of a pattern in data; now for the formal definition. First, fix a computer language—some "friendly" Universal Turing Machine U, which you can imagine as Python.[6] Let len(x) be the *length* of string x, which is just the size of the file in bits. (And len(x, y) is the length of the concatenation of strings x and y, and so on) Then we define:

p is a **pattern** in data x if and only if for some n, $U(p, n) = x$, and $\text{len}(p, n) < \text{len}(x)$.

It is the program portion that encapsulates the regularity of interest—the inputs to that program we can think of as (contextually) unexplained parameters like boundary conditions or error terms. It is worth noting that this notion of regularity is very broad; the "model class" with which we examine the data is *anything computable*. Any pattern extraction algorithm you can think of—such as those employed by modern deep learning systems—is automatically included, but so are many more computable regularities we have yet to dream up. It is hard to picture a model that is *not* computable as being of some use to us.

5. That isn't exactly the program that would work to get the binary file out, but you get the idea. The "hour and a half" estimate is for a bandwidth of 200 Mb/s, which you'd be lucky to have in the US as of 2024.

6. I use "friendly" basically to mean that U should be both *prefix-free* and *additively optimal*. See Li and Vitányi (2008) for details.

A compressible string will generally have many such patterns, one for each such compressing program. Our earlier two programs for producing a trillion ones illustrates this: Roughly translating their pseudo-code above into English, the trillion-ones file has both the pattern "all `1s`" (`for i from 1 to m: print 1`), and *also* the pattern "all the same symbol" (`for i from 1 to m: print j`). The first pattern hardwires the symbol to repeat as part of the regularity, while the second treats it as an input parameter. A string of a trillion *zeros* shares the latter pattern, but not the former, with the string of a trillion ones. And the trillion-ones file also has many sillier patterns, like "substrings with 1,174 of the same symbol in a row, starting at locations l_i, all other symbols from ordered input data d_j."[7] This also counts as a pattern in the file, because it would still losslessly compress the file of a trillion ones. This pattern strikes us as silly because a much *better* compression is obvious for the trillion-ones file, but the "many substrings with 1,174 of the same symbol in a row" pattern would *not* be a trivial compressing pattern for other less uniform strings.[8]

There are three worthwhile mnemonics to keep in mind for this p, n pair read by the universal Turing machine U. First, as we've seen, we can think of them as p as the p-for-Program in the language of U, and n as the n-for-iNput to that program. Second, *only* when together the pair compresses some x, we can also think of them mnemonically as the p-for-Pattern and the n-for-Noise in x—where by "noise" I just mean whatever extra information is needed beyond the particular regularity identified in p to recreate the target string x. Finally, note that the program p effectively encodes a *set* P of strings by its range: $P = \{x \mid \exists n U(p, n) = x\}$. AIT theorists happily call any set to which a string belongs a *property* of that string. In this spirit we can also think of the p as p-for-Property—though strictly speaking, p *encodes* a property by the program that generates it. Then we can think of the n as n-for-iNdex, because it serves as a pointer to pick out a particular member of the set P generated by p.[9] Naturally, this is a very broad conception of "property" that countenances

7. Here we take l_i as a vector of locations to be provided by the input, and d_j as the other symbols filling in the gaps. (If the target string is long enough that "many 1,174-long substrings of the same symbol in a row" is not at all surprising—as would be true of a *very* long string of coinflips—then it would not compress that string, and so not be a pattern for that string.)

8. Ladyman and Ross (2007) would say this silly regularity is not a *real* pattern in the trillion-ones file, because it fails their nonredundancy criterion.

9. A few notes on p as a property: First, the index for elements of the set may not be unique, since different inputs to p may result in the same string. Second, any finite set P is encodable by a program p; in the worst case, simply have p enumerate the members, and output the ith member of the list on input i. Only some infinite sets are encodable by some finite p, however. AIT often restricts its attention to total recursive or even finite sets as its model class, in part because it turns out this is not very restrictive—see Vereshchagin and Vitányi (2004), 3271, section D.

highly heterogeneous sets; the balance of this chapter shows how AIT might identify the more "natural" properties of the type we should call abstractions.

One last preliminary note of importance: p and n are themselves just more binary strings. Where our universe of "things" is information, as represented by binary strings, programs—and so patterns—are just more such "things." This is noteworthy for two reasons. First, it means the pattern and the noise inputs might also have patterns in *them*, which will be important later. Second, if (as I claim) abstractions, universals, or natural kinds are special kinds of patterns, then those too are just more *things*. In this way information-theoretic foundationalism skirts the ancient philosophical ambivalence reflected in the nominalism-realism debates, and thereby (I think) embodies the "mild and intermediate sort of realism" that Dennett initially defended.

4.2 Abstraction and AIT

We are now in a position to see how AIT can help illuminate abstraction. Start with what I take to be a quite generic claim: Abstraction involves *grouping* things, treating them as equivalent or interchangeable for some purpose or another. In other words, it is a kind of coarse-graining that fuzzes out—"abstracts away from"—particular details that are of less interest. Under information-theoretic foundationalism, what this means is that abstractions will be sets of strings.

Already this simple condition on abstraction captures two nearby concepts. First, abstractions will be *multiply realizable*: Any member of the set will instantiate the abstraction.[10] Second, grouping strings together as a set is a standard way in AIT to capture *lossy compression*. Loss*less* compression always reproduces the target string exactly, but the idea of lossy compression suggests that any of a number of instantiations will do equally well for the purpose at hand, and we need not specify the exact details of any one of them. The set of acceptable candidates is the abstraction that the lossy compression targets.

Again, though, not all sets of strings intuitively count as abstractions from their members. I suggested above that the grouping would treat its members as interchangeable "for some purpose," but this sounds suspiciously subjective. It should not be that any set qualifies as an abstraction when I *want* it to be one—that wouldn't even be a "modest" realism about abstractions. My proposal

10. Strictly speaking singletons—sets with a single member—can be abstractions on my account, and so those abstractions will not be *multiply* realizable. But, "in general," they will be, and this is enough to capture the intuitive tie between abstraction and multiple realizability.

is a generalization of Dennett's proposal for mental states: Any abstraction (including mental states) has a real pattern to it.

This needs spelling out, and I think is best motivated by a case study.

4.2.1 A Case Study in Abstraction and a Puzzle

The following case is simple, but I think not too artificial, and it is enough to illustrate what puzzled me for some time about an information-theoretic approach to abstraction. Suppose we are protein biochemists, and we have sequenced a strand of mRNA, recovering a long string of its four chemical bases: adenine, uracil, cytosine, and guanine. Since there are four, we can digitally encode each with two bits, and the result is a digital string twice as long as the number of bases. Three such bases make up a "codon," which encodes for an amino acid (or a "stop" signal). The resulting sequence of amino acids makes a polypeptide, which in turn could make a functional protein—and that is naturally the primary interest for us protein biochemists. Because a few different codons can result in the same amino acid, many different mRNA sequences can make the same polypeptide, as figure 4.1 illustrates.

Since there are twenty different possible amino acids, plus the "stop" signal, we would need five bits to encode each of them.[11] To keep things simpler, though, let's pretend there are exactly sixteen such acids-plus-stop codons, all equally probable, so that we need exactly four bits to encode them. We could convert our digitized mRNA strand into a digitized polypeptide strand, on which our lab will focus its research. A simple program could do this by looking up each six-bit codon encoding and replacing it with its corresponding four-bit amino acid encoding. We thus throw out some information irrelevant to us—namely *which* codon created each amino acid—and save two bits per amino acid in the process. And due to codon redundancy, lots of other mRNA strands would have resulted in exactly the same digitized polypeptide. So we are treating a lot of mRNA strands as the same, for our purposes, and throwing out some irrelevant information to focus on the information we find relevant. It seems clear in this story that we have *abstracted* from our original data.

Now here is what I found puzzling. Suppose that instead of starting with the encoding of an actual mRNA strand, we started with some *random* digital string of equal length as the mRNA encoding, generated by repeatedly tossing a fair coin. We could run this string through the *same* six-bit to four-bit conversion dictionary, treating each six-bit sequence as a codon, and then converting that six-bit sequence into the corresponding four-bit amino acid. The result would be, strictly speaking, a "lossy compression" of the original string

11. Actually we could do it in just under 4.4 bits on average with some cleverness.

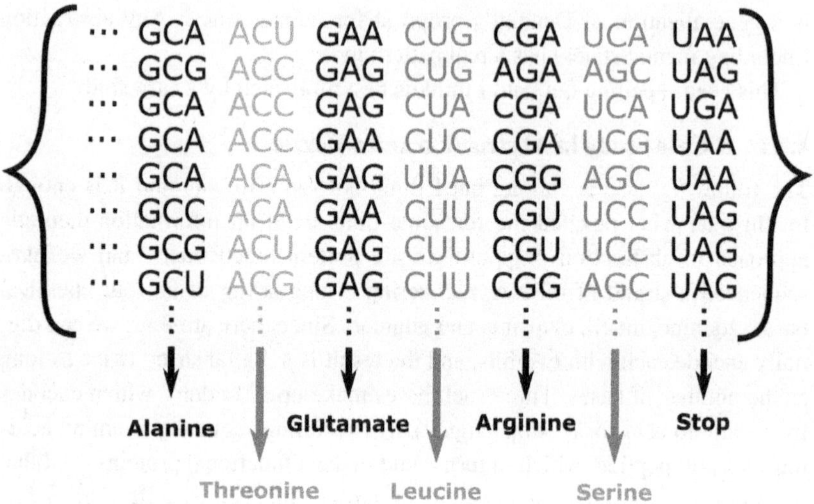

Figure 4.1: mRNA coarse-graining. Reading across each row of letters is a strand of mRNA, as identified by their bases. Reading down the columns of base triplets are codons that all encode for the same amino acid (indicated below each column). The brackets indicate the coarse-graining: These mRNA strands are being grouped as functionally equivalent.

of coinflips. And we could similarly consider the set of all such longer digital strings that would be compressed to that exact same shorter string. I think the set of strings so generated from a random string is *not* an abstraction from the original string. The shorter string does not "throw out irrelevant information" from the coinflips; it's not even clear what that means here. It's just an odd recipe for shortening a long string to a different one of only two-thirds the length. If shortening the coinflip string is *not* abstraction, but shortening the mRNA encoding to the corresponding amino acid sequence *is* abstraction, then the burden is to explain the difference.

Two possible attempts to explain the difference won't work. First, it might be tempting to say that the amino acid strings have *interpretations* (or something similar), while the random string does not. But this will not serve for the desired objective standard. Patternism would like to be able to distinguish the sets of strings on purely information-theoretic grounds, and not by how we use them. Second, the usual AIT instinct to appeal to compressibility to distinguish meaningful and meaningless strings does not obviously work here. The shorter strings generated serve as *lossy* compressions of all their equivalent longer strings, certainly. But lossy compression is cheap—all it takes, in effect, is to throw out some bits. On the other hand, neither the encoded polypeptide nor the shortened random string will *losslessly* compress their original strings, which is the acid test of being a pattern. If I send you only the polypeptide, I have not sent enough information for you to recover the exact mRNA strand that inspired it.

4.2.2 Abstraction Defined

The solution to this puzzle involves shifting attention from ground to figure. We have taken it as a truism that abstraction involves throwing away irrelevant details: For example, we don't typically care about the color of the dress of an options trader when modeling market transactions, and we don't typically care how long the ears are of a particular rabbit in a predator-prey model. What I had missed for a long time is the simple implication that by throwing away irrelevant information, abstraction must *keep* some other information of importance. The patternist instinct can now provide the objective sense of "important information": To keep important information is to *maintain* some pattern in the original data. The amino acid chain does not (losslessly) compress the mRNA chain, but it does maintain a pattern of interest. The amino acid sequences from real mRNA chains—as opposed to randomly generated ones—make viable proteins with biological functions. Such amino sequences are, in fact, compressible thanks to features like repeated "motifs" (short patterns that do specific jobs, like making binding sites) and fairly predictable variation in hydrophobic and hydrophilic sequences (that make the amino acid fold correctly).[12] In the space of possible amino acid sequences, only very few have the properties necessary to fold into a viable protein. In other words, there are patterns in such peptides worth keeping and studying.

Informally, then, an abstraction is just a set of things (i.e., binary strings) that all share some pattern. A bit more formally, a set P counts as an abstraction if all its members are compressible by the same program, given different input parameters. Or, more formally still:

> A set of strings P is an **abstraction** if and only if for some program q, for all $x \in P$, there is an n such that $U(q, n) = x$ and $\text{len}(q, n) < \text{len}(x)$.

Note that the program p that generated the set might itself play the role of q—that is, what generates the set P might *also* compress each of its members. As a toy example, suppose you are coding a 3D virtual world containing many rectangular tables and are tired of hand-coding them. You might write a subroutine that takes as inputs the desired table's location in the world, the

12. See Hosseini et al. (2019) for such a compression tool.

dimensions of its rectangular tabletop, the dimensions of its four legs, and their symmetric offsets from the table edges. This subroutine implicitly generates a set by its inputs—roughly, the set of tables buildable from rectangular prisms. (You would be wise to code limits into the inputs to make sure legs aren't wider than their tabletops, and so on.) Assuming tables must be of a sufficient size that take up a fair number of raw bits to specify, such a subroutine will compress each of its instances, and so the set generator will itself serve as the pattern maintained in the abstraction.

But reflecting again on the mRNA case shows the set generator *p* need *not* itself be a pattern in the set it encodes. That algorithm for converting mRNA encodings to polypeptide encodings just throws out some information we suspect is irrelevant, but tells us nothing about how to compress the data that remains. In fact, the biological amino acid sequences that result *will* be compressible, as we noted. An algorithm to generate all possible source mRNA compatible with the peptide can serve as the set generator *p*, but the regularities of repeated motifs and such serve as the *q*. So although abstracting from mRNA to amino acids in this way will not losslessly compress the original sequencess, such an operation is a successful abstraction because, in fact, it *maintains* patterns that *do* compress losslessly. The lossily compressed coin-flip string, by contrast, will not have such compressing regularities, and so will not have maintained a pattern.

I hope this elaborate setup does not obscure the key, simple ideas: first, abstraction is about grouping together some things that are relevantly similar, thereby eliding their details. Second, the proposed objective standard for "relevantly similar" is that they all share some regularity that enables their compression—that is, they all share a pattern.

4.3 Application: Explanation and Emergence

Now I can sketch how this picture of abstraction might be scientifically enlightening. The bridge between abstraction and science is an algorithmic account of scientific *explanation*, which forms the lynchpin of my patternist project. In brief summary, here are three basic motivations for an account of explanation based in AIT:

- It promises an idealized *standard* for trading a theory's simplicity with its fit to the data, thus formalizing Ockham's razor.[13]

13. This is perhaps best summarized in the "minimum description length" literature; see, e.g., Grünwald (2007).

- It promises to resuscitate the intuitive *unificationist* account of explanation (according to which, roughly, better explanations reduce the number of "brute facts" by displaying the facts' interdependence).[14]
- It comports with the way some recent scientists have theorized about scientific inference.[15]

A fourth potential advantage, explored here, is that when combined with the above account of abstraction, patternist explanation promises a rigorous sense of emergence and higher-level explanations. I will first lay out the briefest overview of patternist explanation, and then show how it connects to patternist abstraction.

Start with the fundamental notion of patternist explanation, a strict standard which I call explanation "proper." The idea of a proper explanation of a given set of explananda is that it extracts *all* and *only* the regularities in the data, leaving the rest as incompressible noise.

> Pattern p^* **properly explains** data x if and only if p^* is a shortest pattern portion of a maximal compression of x. More formally: For some n, $U(p^*, n) = x$, $\text{len}(p^*, n) < \text{len}(x)$, and for any other q, m such that $U(q, m) = x$, both $\text{len}(p^*, n) \leq \text{len}(q, m)$ and $\text{len}(p^*) \leq \text{len}(q)$.

This p^* is what AIT theorists call a "minimal sufficient statistic" of the data, and sometimes called the "meaningful information" of the string. The length of p^* is sometimes called the *sophistication* of the data. This measure contrasts with its (Kolmogorov) complexity, $K_U(x)$, which measures the length of the shortest input to U required to produce x. A long random string, for example, has very high complexity but very low sophistication.[16] (In general I use an asterisk for such special *best* compressions—they get a "gold star.")

In the context where our explananda are something like "everything that is the case," the *proper* explanation will presumably just be the fundamental laws of physics, whatever those turn out to be. In a deterministic universe, for example, from the true fundamental laws plus some unexplained boundary conditions (the "noise") we can, in principle, derive all other phenomena.[17]

14. Perhaps best summarized by Kitcher (1989).

15. For example, the late physicist and mathematician Sir David J. C. MacKay says that one of the themes of his book (2003) is that "data compression and data modelling are one and the same" (31). The physicist Paul Davies (1990) says, "Given some data set, the job of the scientist is to find a suitable compression, which expresses the causal linkages involved" (63). And the physicist Max Tegmark (2014) says, "There are several science definitions that I like, and one of them is *data compression*, explaining a lot with a little" (chapter 7, section 9).

16. See, e.g., Vitányi (2006) or Mota et al. (2013).

17. Indeed AIT provides a promising foundation for a Lewis-style "Best System Analysis" of scientific laws, as Torza (2022) suggests.

This is an extremely demanding condition on an explanation, and it serves only as an ideal that we will not achieve in practice. (For one thing, finding such a proper explanation is uncomputable.) We thus need to supplement this account, to characterize better and worse approximations of this ideal. Current scientific laws as we understand them are not the complete story, surely—but they are a better story than Newton's. To capture such imperfect explanations, patternism can use what James Woodward (2003) calls the "hidden structure strategy," following Peter Railton's (1981) proposal that there is an "ideal explanatory text" about which approximate or pragmatic explanations provide nonideal information. In the patternist framework, the defined proper explanation plays the role of the ideal explanatory text, and AIT allows us to define formally what it is for some *partial* explanation to "provide information" about the proper explanation.

First, in AIT, string *a provides information* about another string *b* just in case $K_U(b \mid a) < K_U(b)$, where $K_U(b \mid a)$ is the *conditional Kolmogorov complexity*: The length of the shortest (p, n) required to produce *b* given string *a* as input "for free." (This, of course, is intended to be analogous with conditional probabilities.) In sum, *a* provides information in this sense about *b* when *b* is easier to compress if *a* is already known.[18]

With this machinery in place, the "hidden structure strategy" is easy: We can define a *partial* explanation as a pattern that provides information about the properly explaining pattern.[19]

> Pattern *p* **partially explains** data *x* if and only if *p* provides information about *x*'s proper explanation p^*. Formally, for some n, $U(p, n) = x$, and $\text{len}(p, n) < \text{len}(x)$, and $K_U(p^* \mid p) < K_U(p^*)$.

Now we are in a position to apply our definition of abstraction, in order to characterize explanations in the special sciences that do not advert to fundamental physical laws—in other words, explanations that work *only* at a higher level of abstraction. I call these explanations "parochial." (Since I couldn't easily avoid all the alliteration, I decided to commit to it instead.) Again we start with a "proper" such explanation:

18. The measure of *how much* easier, in bits, is called the *algorithmic mutual information* between *a* and *b*: $I(a:b) = K_U(b) - K_U(b \mid a)$. This is intended to be analogous with the more traditional mutual information from Shannonian information theory. The "mutual" is justified in both cases because this relation is symmetric—or more carefully, in the algorithmic case, it is symmetric up to a constant, once defined a bit more carefully. See Grünwald and Vitányi (2003) section 5.2.

19. Note that, strictly speaking, this definition counts the proper explanation as partial. Given the convention in mereology to call wholes "parts," I think it's not so bad also to call the whole explanation a partial one—though I suppose it commits me to saying that proper explanations are not *properly* partial.

> Pattern p'^* **properly parochially explains** abstraction P if and only if it is a proper explanation of that abstraction. Somewhat more formally: p'^* extracts all patterns shared by elements of P. Still more formally: p'^* compresses all elements of P, and there is no shorter program that does so.

In general, I use a prime marker for higher-level patterns. And from this we can get, finally:

> Pattern p' **partially parochially explains** abstraction P if and only if p' is a partial explanation of that abstraction. That is, p' compresses all elements of P, and provides some algorithmic information about its proper explanation p'^*.

I recognize this is a lot of definitions and notation, but the underlying idea is simple: Abstraction allows us to treat coarse-grained data as the target of explanations, independently of its various realizers. And although a parochial explanation will always represent a *pattern* in any underlying realizer x of abstraction P, the parochial explanation may not be an *explanation* of those underlying data—not even partially. Often the pattern is only explanatory at the abstract level. That is what makes the pattern a "parochial" explanation, and I think this is how we want it. For example, $PV = nRT$ helps summarize a great deal of information about the behavior of any "gas" (an abstraction), and thus will help compress the micro-level facts, independently of whether the correct laws governing the fundamental particles underneath are quantum or classical. Similarly, a genuine pattern in predator-prey relationships will also give us *no* insight into the fundamental laws of physics. Finally, consider a fun, clean example that harkens back to Dennett's seminal paper: We can make a Game of Life cellular automaton on top of a *different*, micro-cellular automaton underneath.[20] Able to see only the macro-level cells, we could glean genuine patterns in their dynamics (such as that they follow John Conway's original rules) without thereby gaining *any* information about the micro-level rules that determine the evolution of its cells. The rules at the fundamental level (together with the initial conditions) will surely be the most compact summary of the full state at any time, but this does not, in my view, make the patterns in the higher-level phenomena any less explanatory of the behavior at that higher level.

Finally, I want to suggest briefly that explanations functioning only at the level of an abstraction point the way toward a legible account of *emergence*. Following roughly the Quinean dictum that "to be is to be the value of a variable" in a best theory (Quine 1948), and supposing that we cannot dispense

20. Thanks to my coeditor Tyler Millhouse for this example. See https://www.youtube.com/watch?v=xP5-iIeKXE8 for a dramatic illustration of one CA on top of another. In the video both levels probably follow Conway's rules, but since there are many Turing-complete CAs, this needn't be the case.

with explanations at the level of abstracta (if only because we have neither the computational nor observational capacities to be Laplacean demons), our best theories will posit legitimate things at abstract levels. Basically, I propose that if something features in a best explanation at a coarse-grained level, it's emergent from the lower level.

Patternism even countenances talk about *causation* at the abstracted level, if we are sufficiently Humean-Lewisian about it. Think again of the metaphysical possibility of the simulation hypothesis: If in fact a digital simulation is under all the phenomena of the physicists' standard model, it still can be the case that electrons exist, and *cause* other electrons to veer away at that level of abstraction. They are, as patterns in the underlying data, still real enough.

4.4 Conclusion

I have come to think this algorithmic take on abstraction is both rigorous and intuitive. Though the details are technical, the main idea again is simple: An abstraction is a grouping of particulars that share an important similarity, and differ only with regard to details in realizing that similarity. The proposed objective standard for the "important similarity" is that the information of the instances all must share a pattern—which is to say, a regularity that allows for compression. This algorithmic construal of abstraction can help clarify debates in realism, explanation, and emergence as part of a larger "patternist" package.

References

Bird, A. 2023. "Restricted Composition Is Information Compression." *The Philosophical Quarterly* 73 (3): 677–700. ISSN: 0031-8094. https : //doi.org/10.1093/pq /pqad034.

Bostrom, N. 2003. "Are You Living in a Computer Simulation?" *Philosophical Quarterly* 53 (211): 243–255.

Chalmers, D. J. 2005. "The Matrix as Metaphysics." In *Philosophers Explore the Matrix,* edited by C. Grau, 132–176. Oxford University Press.

Chalmers, D. J. 2022. *Reality+: Virtual Worlds and the Problems of Philosophy.* W. W. Norton.

Davies, P.C.W. 1990. "Why Is the Physical World So Comprehensible?" In *Complexity, Entropy, and the Physics of Information,* edited by Wojciech H. Zurek, 8:61–70. Studies in the Sciences of Complexity. Santa Fe Institute, Addison-Wesley Publishing Company.

Dennett, D. C. 1991. "Real Patterns." *Journal of Philosophy* 88 (1): 27–51.

Grünwald, P. D. 2007. *The Minimum Description Length Principle.* MIT Press.

Grünwald, P. D., and P.M.B. Vitányi. 2003. "Kolmogorov Complexity and Information Theory." *Journal of Logic, Language, and Information* 12:497–529.

Hosseini, M., D. Pratas, and A. J. Pinho. 2019. "AC: A Compression Tool for Amino Acid Sequences." *Interdisciplinary Sciences: Computational Life Sciences* 11 (1): 68–76. ISSN: 1867-1462. https://doi.org/10.1007/s12539-019-00322-1.

Kitcher, P. 1989. "Explanatory Unification and the Causal Structure of the World." In *Scientific Explanation,* edited by P. Kitcher and W. C. Salmon, XIII:410–505. Minnesota Studies in the Philosophy of Science. University of Minnesota Press.

Korman, D. Z. 2011. "Ordinary Objects." In *The Stanford Encyclopedia of Philosophy,* Winter 2011, edited by E. N. Zalta. Metaphysics Research Lab, Stanford University.

Ladyman, J., and D. Ross. 2007. *Every Thing Must Go.* Oxford University Press.

Lewis, D. 1986. *On the Plurality of Worlds.* Blackwell.

Li, M., and P.M.B. Vitányi. 2008. *An Introduction to Kolmogorov Complexity and Its Applications.* Springer.

MacKay, D.J.C. 2003. *Information Theory, Inference, and Learning Algorithms.* 2004 edition. Cambridge University Press.

Mota, F., S. Aaronson, Luís Antunes, and André Souto. 2013. "Sophistication as Randomness Deficiency." In *Descriptional Complexity of Formal Systems,* edited by H. Jurgensen and Rogério Reis, 172–181. Springer Berlin Heidelberg. ISBN: 978-3-642-39310-5.

Petersen, S. 2019. "Composition as Pattern." *Philosophical Studies* 176 (5): 1119–1139. ISSN: 1573-0883. https://doi.org/10.1007/s11098-018-1050-6.

Quine, W.V.O. 1948. "On What There Is." *Review of Metaphysics* 2 (5): 21–38. ISSN: 00346632. http://www.jstor.org/stable/20123117.

Railton, Peter. 1981. "Probability, Explanation, and Information." *Synthese* 48:233–256.

Tegmark, M. 2014. *Our Mathematical Universe: My Quest for the Ultimate Nature of Reality.* EPUB edition. Alfred A. Knopf.

Torza, A. 2022. "Laws of Nature and Theory Choice." *Synthese* 200 (6). ISSN: 1573-0964. https://doi.org/10.1007/s11229-022-03950-1.

van Inwagen, P. 1990. *Material Beings.* 1995 edition. Cornell University Press.

Vereshchagin, N. K., and P.M.B. Vitányi. 2004. "Kolmogorov's Structure Functions and Model Selection." *IEEE Transactions on Information Theory* 50 (12): 3265–3290. https://doi.org/10.1109/TIT.2004.838346.

Vitányi, P.M.B. 2006. "Meaningful Information." *IEEE Transactions on Information Theory* 52 (10): 4617–4626.

Woodward, J. 2003. *Making Things Happen: A Theory of Causal Explanation.* EPUB edition. Oxford University Press.

Zeilinger, A. 2004. "Why the Quantum? 'It' from 'Bit'? A participatory universe? Three far-reaching challenges from John Archibald Wheeler and their relation to experiment." In *Science and Ultimate Reality: Quantum Theory, Cosmology, and Complexity,* edited by J. D. Barrow, Paul C. W. Davies, and C. L. Harper, Jr, 201–220. Cambridge University Press.

5 The Problem of Platonic Codes

Tyler Millhouse

5.1 Introduction

Across a wide body of work, Dennett proposes three *stances* from which we might predict the behavior of human beings and other complicated systems (see, e.g., Dennett 1971, 1989, 2009). The *physical stance*, *design stance*, and *intentional stance* involve distinct assumptions and, where successful, distinct ontological commitments. For Dennett, a stance is successfully applied when it allows us to "usefully and voluminously" predict the behavior of the target system (2009, 339). The successful application of a stance, in turn, depends on the existence of *real patterns* within the target system (Dennett 1991). For the intentional stance, these patterns ground the mental states ascribed to the system by those taking the stance (e.g., beliefs and desires).

Of course, success comes in degrees, and so too do the patterns on which the success of a stance depends. Further, these patterns may not be "the direct outline of a concrete orderly process in the background" (e.g., a physical symbol system that represents attitudes and their contents in a language of thought; Dennett 1991, 43). This makes Dennett's realism about the propositional attitudes a "mild realism," at once recognizing the distinctive power of intentional stance prediction and avoiding "industrial-strength" commitments about how propositional attitudes are instantiated (1991, 30).

In this chapter, I will bracket debates about whether Dennett's realism is too weak or too strong—though I have commented on this elsewhere (Millhouse 2021, 2022). Instead, I will provide a more fundamental information theoretic motivation for the kind of stance-taking Dennett advocates, a motivation that bolsters the connection between stance-taking and ontology. For this purpose, I will consider the physical, design, and intentional stances collectively as distinct modes of the *predictive stance*. The predictive stance is taken by agents interested in predicting events in their environment, especially events that would result from possible actions.

I will contrast this stance with the *descriptive stance*. The descriptive stance is taken by agents interested in efficiently representing events in their environment for the purpose of communication. While all stances are taken by agents, the predictive stance depends much more on the details of the stance-taking agent. The kinds of predictions that serve an agent's interests will vary with their goals and their means for achieving them (e.g., their sensors and effectors). The descriptive stance, in contrast, assumes a narrowly specified goal and can be approached as a formal problem in coding theory.

As I will argue, it is *only* when we take the predictive stance that the world resolves into anything remotely recognizable (ontologically speaking). Nevertheless, while there are overwhelming practical reasons to take the predictive stance, there are no obvious theoretical reasons to prefer it over the descriptive stance as an approach to faithfully representing one's environment. If this is correct, then we must accept that any attempt to decouple our ontology from either stance-taking or our practical aims as agents is unlikely to succeed.

At the same time, my argument in no way suggests that the successful application of a predictive stance is a subjective matter. As Dennett argues elsewhere:

> My thesis will be that while belief is a perfectly objective phenomenon (that apparently makes me a realist), it can be discerned only from the point of view of one who adopts a certain *predictive strategy*, and its existence can be confirmed only by an assessment of the success of that strategy (that apparently makes me an interpretationist). (1989, 15; emphasis in original)

My argument, then, is consistent with Dennett's commitments to both realism *and* interpretationism. It holds that the propositional attitudes (or virtually any other element of our ontology) can be objectively confirmed only by successfully adopting one kind of predictive stance or other. This success, in turn, depends on real patterns in the target system.

That said, the details of this argument may be surprising to many of those sympathetic to real patterns. For example, while Dennett (1991) relies on the analogy between predictive modeling and compression, I most closely associate efficient (i.e., compressed) representations of events with the descriptive stance. Moreover, I will show that any successful application of the predictive stance will involve a language for representing events that purposefully avoids maximizing efficiency. This point is critical for appreciating (1) how the predictive stance differs from the descriptive stance and (2) the role of an agent's goals in fixing an appropriate language for representing events. Crucially, none of this undermines the connection between successful stance-taking and the existence of real patterns in the target system.

To make my argument, I will proceed as follows. First, I will introduce an attempt by Plato to put taxonomy on an objective footing, which proposes (1) a normative constraint on taxonomy and (2) a sketch of an algorithm for categorizing things in compliance with this constraint. Second, I will argue that despite its superficial implausibility, Plato's proposal anticipated central elements of information theory. Third, I will argue that what I call *Platonic codes* are not a suitable language for prediction *precisely because* they are maximally efficient languages for description. In brief, Platonic codes obscure the kinds of patterns agents must represent in order to successfully predict how their actions will affect the world. As a result, agents must choose between optimal languages for description and optimal languages for prediction, and it is only in languages of the latter kind that the familiar world emerges. Instead of attempting to resolve this dilemma, I will conclude by proposing that neither type of language is preferable over the other *except* for practical reasons.

5.2 Platonic Codes

Across multiple dialogues, Plato proposes a norm of taxonomy that requires "dividing things . . . by classes, where the natural joints are, and not trying to break any part, after the manner of a bad carver" (*Phaedrus*, 265e). Here and elsewhere, Plato is concerned with the question of *natural kinds*, and he is especially worried that commonsense taxonomies represent our parochial interests more than joints in nature (e.g., *Statesman*, 263d). To address this problem and to discipline our thinking as we approach taxonomy, Plato proposes a curious method for dividing things into classes. Wherever a plausible basis for a distinction can be found, we should divide the members of each parent class into two equal subclasses.[1] For example, Plato argues that the human/beast distinction is suspect because it isolates a small subset (humans) of the parent category (animal) and distinguishes it from a much larger subset (beasts). The distinctions between animals and other living things is in better standing because it more closely approximates an equal binary division. The upshot is that our process of classification should *ideally* result in a nested hierarchy where each branching point divides its parent branch into two equally populated daughter branches.

1. Plato describes this approach as follows: "It was as if a man should think he was dividing number into two classes by cutting off [ten thousand] from all the other numbers, with the notion that he was making one separate class and then should give one name to all the rest, and because of that name should think that this also formed one class distinct from the other. A better division, more truly classified and more equal, would be made by dividing number into odd and even, and the human race into male and female; as for the Lydians and Phrygians and various others they could be opposed to the rest and split off from them when it was impossible to find and separate two parts each of which formed a class" (*Statesman*, 262d–263a).

This is certainly a strange proposal, and one would be forgiven for dismissing it out of hand. It does have plausible implications in some cases (e.g., that the Greek/barbarian distinction is rather tenuous), but the insistence on equal binary divisions seems wildly unmotivated. It *is* reasonably interest-neutral as criteria go, but that could be said about a wide range of arbitrary methods of division. Why not favor 60-40 binary divisions or 25-25-25-25 quaternary divisions? Neither of those seem to align with any particular set of interests. Nevertheless, a proposal quite similar to Plato's plays a central role in modern coding theory. As we will see, consideration of this role actually undermines his proposal in certain use cases, albeit for much subtler and more definitive reasons than those just discussed.

Suppose we have a set of mutually exclusive events that we wish to communicate to a partner. A single event happens for each unit time, and we will communicate which event has happened at each moment by sending a message. We can confer with our partner in advance and agree on what message we will use for which events, but we will be separated when the events occur and will have to rely on our preestablished code for communication. By creating this code, we hope to minimize the average length of the messages we must send. For simplicity, I will assume that we intend to communicate using zeros and ones (i.e., using a binary code).[2] What properties must this code have? More specifically, how should we assign strings of zeros and ones (i.e., messages) to events such that we can minimize the average length of our messages over time?

Fano's (1949) answer to this question (hereafter, "Fano coding") is one that Plato would appreciate. Instead of repeatedly dividing the set of possible events into two subsets of equal *size*, we should repeatedly divide the set of possible events into two subsets of equal *probability*. To accomplish this, events are listed from most to least likely. This list is then divided at whichever index maximizes the equiprobability of the events that come before and the events that come after. The resulting subsets are subject to the same sorting and division process whenever they contain at least two events. Sorting and division stop once all subsets contain exactly one event. We can interpret this process as producing a branching tree, whose root is the set of all possible events and whose leaves are singleton sets containing exactly one event. As illustrated

2. This assumption might seem like it grants too much to Plato, but, as we will see, the connection between Plato's proposal and coding theory lies in the equality of divisions irrespective of whether those divisions are binary, trinary, quaternary, and so on. Similarly, the problem for Plato's approach does not depend on whether our chosen code is binary, trinary, quaternary, and so on.

in figure 5.1, each branching point of this tree divides the parent set into two daughter subsets.

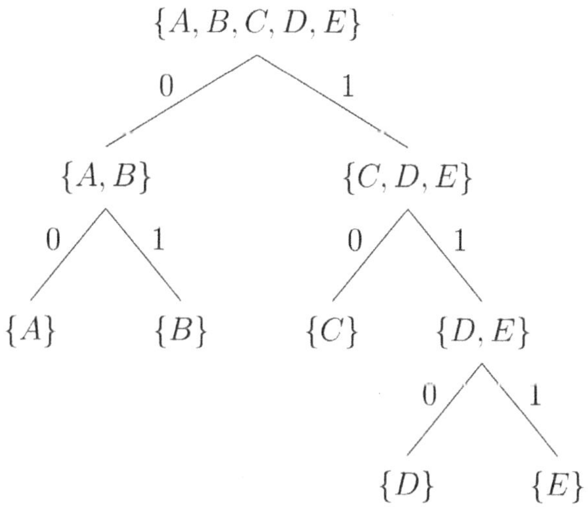

Figure 5.1: A code tree generated by Fano coding given events $\{A, B, C, D, E\}$ with probabilities $\{0.4, 0.15, 0.15, 0.15, 0.15\}$, respectively.

Given such a tree, we can create "directions" in binary for reaching any leaf node from the root node. At each branching point, we will indicate which of the two branches to follow using either a "0" or a "1". We will only use as many directions as necessary to reach the desired leaf node, and each leaf node will have a unique set of directions. Thus, these "directions" can be used as a message, encoded in binary, to uniquely indicate which event has occurred. In addition, this tree structure means that no sequence of directions for one leaf will be a *prefix* of any other leaf's sequence since each leaf terminates a branch. For example, if the sequence for event *e* is "01011", no other event will have a sequence beginning "01011 . . ." This means that no spaces or other delimiters are required to communicate multiple messages. Once a leaf (event) is reached, we can return immediately to the root and await the next direction sequence.

This explains how Fano coding generates codes for events, but it doesn't motivate the underlying approach (i.e., repeatedly divide the set of events into subsets of equal probability). Within the constraints of the sorting and division process, Fano coding seeks to maximize the information content of each bit of a message one bit at a time starting with the first bit. Here, the amount

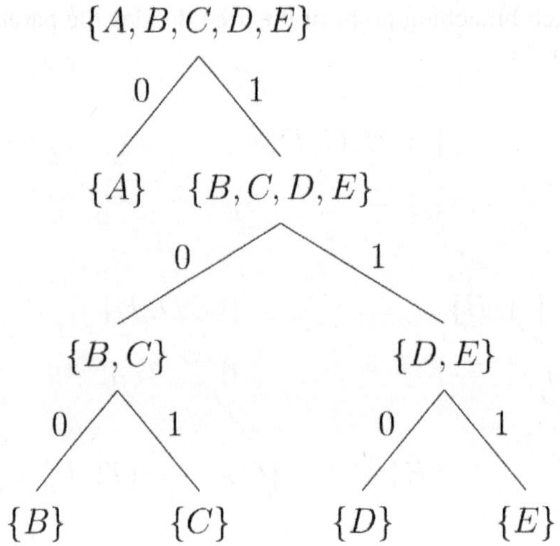

Figure 5.2: A code tree generated by Huffman coding given events $\{A, B, C, D, E\}$ with probabilities $\{.4, .15, .15, .15, .15\}$, respectively.

of information carried by a bit corresponds to how much that bit causes our partner to change the prior probabilities they have assigned to events. Learning that an unlikely event has occurred or a likely event has not occurred will induce a large update in our partner's probabilities. Learning that a likely event has occurred or an unlikely event has not occurred will induce a small update in our partner's probabilities.

It is hard to say exactly what this tells us about how we should use bits to communicate efficiently. For example, suppose a zero tells our partner that a very unlikely event has occurred. On the one hand, this will induce a large change in their probabilities. On the other, the unlikely event (and the message communicating it) will occur only rarely. As it happens, a single bit will do the most (on average) to change our partner's probabilities if it tells them which of two equiprobable events has occurred. I have given a general proof of this result in the appendix, but we can easily see that this result holds in a particular case. Suppose there is a 50-50 chance that *A* will (or will not) occur. In this case, using a single bit to communicate whether A happened will always change our partner's confidence in *A* by 50%. Suppose instead that there is a 25-75 chance that *A* will occur. In this case, 25% of the time the bit will change their confidence by 75%, and 75% of the time it will change their confidence by

25%. In the former case, the average change will be 50%, but in the latter case, the average change will be only 37.5%.

Of course, it may not be possible to divide every relevant set of events into two equiprobable subsets. For example, suppose we only want to tell our partner whether or not it is raining, and we send a message updating them on the weather every half hour. If we live in an area where it rains 50% the time, the messages will be highly informative. If we live in an area where it rains only 1% of the time, then our messages will be considerably less informative. Irrespective of the chances of rain, however, we can do no better than to assign "0" to one state of the weather and "1" to the other. Such a code will be optimal in the sense that there is no better code, but it will not be ideal in the sense that different event probabilities would permit more informative messages.[3]

The possibility of recalcitrant probabilities also exposes a weakness of Fano's approach that makes it less than optimal. Notice that Fano's algorithm is "greedy" in that it maximizes the information content of the current bit without regard for future bits. This greedy algorithm works well but is not optimal. For example, suppose we want to communicate five events, A, B, C, D, and E with probabilities 0.4, 0.15, 0.15, 0.15, and 0.15, respectively. Consider the following code generated by Fano's algorithm: A = "00", B = "01", C = "10", D = "110", E = "111". Compare this to a code generated by Huffman's algorithm (Huffman 1952): A = "0", B = "100", C = "101", D = "110", E = "111". Multiplying the probability of each message by its message length, we find the expected message lengths under each encoding:

$$0.4 \times 2 + 0.15 \times 2 + 0.15 \times 2 + 0.15 \times 3 + 0.15 \times 3 = 2.3 \qquad (5.1)$$

$$0.4 \times 1 + 0.15 \times 3 + 0.15 \times 3 + 0.15 \times 3 + 0.15 \times 3 = 2.2 \qquad (5.2)$$

The reason that Huffman coding achieves this modest economy is not important here, but it *is* provably optimal, even for nonideal probabilities. That said, it is worth emphasizing that Huffman coding doesn't undermine the idea that a single bit of code will ideally tell us about which of two equiprobable events has occurred. It is simply that different algorithms are required to generalize this idea to longer codes and less-than-ideal probabilities.[4]

3. Plato also anticipated this kind of problem (see Fn. 1).

4. Huffman coding (1952) works backward, starting with the last bit of any code and working backwards to the first. Equivalently, instead of building a code tree from root to leaves, it builds it from leaves to root. Imagine we start with a list of singleton sets sorted by their probability, each containing one of our events. Huffman coding takes the two *least* probable sets and makes a parent set with both sets as members. It then removes the two daughter sets from the list and adds the parent set to the list, setting its probability to the sum of the probabilities of its children. Next, it sorts the list by probability and repeats this process until the list contains a single set with all events as members of some subset or other. The nested set/subset structure corresponds to a binary tree,

However, what (generally speaking) makes a set of event probabilities ideal? Ideal probabilities have values that can be expressed in the form $\frac{1}{2^i}$ or (equivalently) 2^{-i}, where i is a whole number. These values are those that result from starting at one and dividing by two i times. Since the probabilities for all events must sum to one, events with ideal values will always allow us to perfectly divide sets or subsets of events into two equiprobable subsets via Fano's algorithm. What I have called "ideal probabilities" are also known as "dyadic distributions" (Cover and Thomas 2006, 151). Cases with ideal probabilities allow us to understand an important general result in information theory. Given ideal probabilities and a maximally efficient code, the length of a message, m, encoding an event, e, is given by the following equation:

$$|m| = -\log_2(P(e)). \tag{5.3}$$

This equation simply tells us how many equiprobable binary divisions are required to specify an event with a given probability. An event with probability 0.5 requires only one bit, an event with probability 0.25 requires two bits, and so on. Hence, the expected length of messages sent, L, where each message encodes an event in our event set, $e_n \in \mathbb{E}$, is given by the following equation:

$$L = -\sum_{0}^{N} P(e_n)\log_2(P(e_n)). \tag{5.4}$$

Not coincidentally, this equation is simply the entropy of a random variable (in units of bits) whose outcomes are the events in our event set (Shannon 1948). Entropy is a measure of randomness. It is minimized when only one outcome ever occurs (i.e., has probability one), and it is maximized when all outcomes are equiprobable (i.e., have a uniform distribution). This means that the expected message length under an optimal code for events with ideal probabilities will directly reflect the degree of randomness in the events the messages encode. More random events require longer codes, while more regular events allow shorter codes. This result is one way of understanding why efficient representation depends on regularities in data (i.e., the idea that motivates real patterns; Dennett 1991).

and codes are assigned in the same manner as before, providing directions from root to leaf for each event. The intuition here is that if nonideal probabilities do force us to use an extra bit here or there to distinguish messages, then it would best used to distinguish the least likely messages (i.e., where it is least likely to be sent).

5.3 Separating the Chaff from the Wheat

My aim in this section is to demonstrate how coding algorithms attempt to distill the randomness and unpredictability of the real world into a purified language, one that contains information about the order and regularity of the world only implicitly (i.e., in the probabilities of events represented by code lengths and in our choice of events). I will give a concrete example of this distillation process presently, but it is worth pausing to clarify some terminology.

Character	Frequency
z	880
q	2,150
x	2,978
j	3,164
k	9,588
v	17,786
b	28,078
g	32,454
p	32,606
y	37,118
w	40,314
f	44,422
m	44,800
c	48,286
u	53,630
d	70,932
l	72,078
r	109,326
s	122,742
h	125,376
n	133,498
i	137,446
o	143,614
a	149,202
t	177,884
e	241,640
_	432,569

Table 5.1: The frequencies of characters in Plato's *Republic* including only letters (ignoring case) and spaces.

For present purposes, I will use the term "optimal codes" to refer to codes that minimize expected message length given a set of event probabilities—irrespective of whether those probabilities are ideal. I will use the term "ideal code" to refer to optimal codes with ideal event probabilities. In actual practice, very few bits are wasted due to less-than-ideal probabilities. Huffman code lengths will not exceed their ideal values (given by equation 3) by even a full bit. However, this does not mean that most codes are even approximately as efficient as they might be. Care is required here since we actually *can* create codes that are approximately as efficient as possible *given* an assignment of probabilities to a chosen set of events (e.g., by using Huffman coding). It is much harder to create a code that is maximally efficient for a *correct* assignment of probabilities to a *properly* chosen set of events. I will reserve the honorific "Platonic" for ideal codes that are maximally efficient in the latter sense.

A fairly straightforward consequence of constructing a Platonic code is that messages sent using such a code do not look like messages at all. By this, I mean that if we were to spy on two people using an unknown Platonic code, we could not distinguish the intercepted messages from random noise. Moreover, even the recipient of the encoded messages (who knows how to decode them) could not predict which bit will come next better than chance. To see why this must be the case, note first that every bit sent is either the first bit of a message or a subsequent bit of a message. As we saw, the first bit of any message should tell our partner which of two equiprobable events (or disjunctions of events) has occurred, and every subsequent bit is equiprobable given the preceding bit. This means that irrespective of the bits they have seen so far, our partner should always place a probability of 0.5 on the next bit being a 1 (or a 0). Hence, a sequence of bits generated by partners communicating via a Platonic code should be indistinguishable from a sequence of bits generated by fair coin flips.

As an exercise, let's put this idea to the test using an English translation of Plato's *Republic*. Restricting our attention to spaces and letters (ignoring case), we find the character frequencies shown in table 5.1. Converting these frequencies to probabilities, we can compute the entropy of character probabilities using (4). The entropy is 4.06 bits. We can also check the average message length achieved by Huffman coding. The average message length is 4.10 bits. Unsurprisingly, there is close agreement between the average message length under an optimal code and the entropy of the event probabilities.

Nevertheless, it is not yet clear why we should expect later bits to be unpredictable given earlier bits. That would be true (to a very close approximation) if we looked at data that was actually generated by sampling characters purely based on their individual probabilities. Nevertheless, the probability of seeing a particular English letter depends heavily on which letters came before. A "g" is more likely to follow an "n" than a "p", and hence, the code for "g" is more likely to follow the code for "n" than the code for "p". Consider the consequences of this for individual bits. Let's suppose that the last code was for an "n" and that the codes for "g" and "p" are distinguished by only their final bit. This is entirely plausible under the coding schemes discussed earlier since "g" and "p" have very similar frequencies and, thus, the final bit distinguishing "g" and "p" seems to tell us about which of two very nearly equiprobable events has occurred (table 5.1). However, in this case the bit no longer tells us about two equiprobable events since "ng" appears 12,744 times in the *Republic* while "np" occurs only 60 times, making the former over 200 times more likely than the latter. Note that this problem arises due to the *dependence* of later characters on earlier characters and not the nonideality of the character probabilities.

For this reason, unless individual character probabilities are independent of each other, a code constructed using only individual character probabilities cannot be maximally efficient (in the Platonic sense), since individual bits will fail to tell us about equiprobable events. This is true even if the code is optimal or ideal given our probability assignments. As I noted earlier, most inefficiencies do not arise from nonideal probabilities. Instead, they arise from our ignorance of the full joint probability distribution over the relevant elementary events (i.e., the probability distribution over all possible sequences of characters).

If each character in a sequence really were independent of those that came before, then the individual character probabilities would be sufficient to specify a joint distribution over all character sequences, where the probability of any sequence would be equal to the product of the probabilities of each character it contains. However, this independence assumption is not satisfied for any natural language, and our ignorance of the full joint distribution results from the difficulty of characterizing longer-range dependencies. For example, we know that the words in the first chapter of a book constrain the words in the last chapter *probabilistically*, but it is difficult to imagine how we might formalize that dependence in terms of individual characters for the purpose of optimizing our code.

That said, there very much *are* ways of incorporating dependence relationships between events where we have the tools to formalize them. I will illustrate two simple strategies here—both of which require us to abandon individual characters as the events we wish to encode. First, we might switch from encoding individual characters to encoding ordered pairs of characters (or "two-grams"). This will allow us to assign a (much) longer code to "np" than to "ng." As before, we can estimate the benefits of this strategy by computing the expected message length under an ideal code as the entropy of the two-gram probabilities in the *Republic*. The entropy is 7.31 bits.

Crucially, we should still expect a reduction in the number of bits required to encode the full text even though the expected code length for a character-based encoding was only 4.06 bits, since there are half as many two-grams in the *Republic* as there are characters. For example, the bits required to encode two separate characters under an ideal character-based encoding would be 8.12 (4.06×2)—a notable increase over 7.31. If an ideal character-based encoding requires 8.12 bits for two characters and an ideal two-gram-based encoding requires 7.31 bits per two-gram, we should expect a roughly 9.98% reduction in the length of the full text encoding. Using the figures shown in table 5.2, we see that the actual reduction is slightly higher at 10.55%.[5]

5. This is likely due to the fact that codes with a larger number of lower probability events (like the two-gram code) have an easier time dividing events into equiprobable subsets, so the two-gram code is probably closer to ideal than the character code. Indeed, the average message length is

Events	Bits
Uniform	11,572,805
Characters	9,495,153
2-Grams	8,493,046
3-Grams	7,635,067
4-Grams	6,858,575
Words	4,436,342

Table 5.2: Bits required to encode Plato's *Republic* under different representations of events. "Uniform" encodes characters as if they were equally likely and disregards their observed frequencies, minimizing compressibility.

We can easily extend this approach by considering three-grams, four-grams, and so-on. As we might expect, this does result in shorter encodings for the full text (table 5.2) as we exploit longer and longer range dependencies to optimize our code. In addition to the *n*-gram strategy, we might use our domain knowledge of the English language to parse the text into words and assign codes to words. The actual domain knowledge required given this preprocessed text is minimal (i.e., that words are separated by spaces). This is a more natural approach and results in an even larger reduction in the length of the encoded text (table 5.2). Similar strategies might be used whenever we have this kind of domain-specific knowledge. In addition, these strategies can easily be used outside the domain of language (see, e.g., intraframe vs. interframe compression for video).

What matters is that these each of these strategies bring more and more of our knowledge of character sequence probabilities to bear in optimizing our code. In cases where our knowledge is more complete, we can achieve more efficient codes. Further, as we bring more and more of our knowledge to bear in optimizing our code, we more closely approximate a Platonic code, and the encoded text will more closely resemble uniform random bits. In the limit, we will select events (e.g., characters, *n*-grams, or words) which allow us to incorporate *all* of the inter-character dependencies implicit in our full joint distribution over character sequences. Neither these events nor the bits that encode them will predict any future events or bits. The code that results will communicate all and only those bits needed to resolve our residual uncertainty about how any particular sequence will proceed given what has come before.

5.4 The Predictive Stance

Having carefully laid out these information-theoretic results, the main conclusion follows almost immediately. Where Platonic codes are concerned, no bit

about 101.0% of the entropy for the character code (4.10 bits vs. 4.06 bits) and about 100.4% of the entropy for the two-gram code (7.34 bits vs. 7.31 bits).

can be predicted from *any* sequence of earlier bits—no matter how comprehensive. Where we have accurate (but not ideal) probabilities for optimally chosen events, our ability to predict any bit given any sequence of earlier bits will be minimal and related primarily to the constraints of a binary code. Where we have tried our best but have inaccurate probabilities or suboptimal events, our ability to predict any bit given any sequence of earlier bits will be minimal since the knowledge required to make accurate predictions is the very knowledge we lacked when optimizing our code. Hence, while the descriptive stance involves seeking a maximally efficient code, the predictive stance will *of practical necessity* require a code that departs from maximal efficiency.

To be more specific, a predictively useful code will not exploit (for the purpose of enhancing efficiency) regularities relating events the agent can influence to events that constitute or influence the satisfaction of the agent's goals. For example, imagine a stream of characters are being generated by a Markov chain (i.e., where probabilities for the next character depend only on the current character). Suppose there is an agent that can either take no action or intervene in this process by taking one of two actions. These actions alter the next character probabilities. The agent's goal is to cause the process to produce the letter "p" as often as possible. The agent is limited in that, regardless of the current state (i.e., character), no action results in a high probability of "p". Further, the most likely character to result from doing an action in a state never has a high probability of transitioning to "p". The result is that the agent must learn the state transitions that take place without intervention and through occasional interventions guide the process to states that have a high probability of reaching "p" after a few transitions.

How would this agent fare under an *n*-gram encoding of the states (i.e., characters)? First, only nonoverlapping sequences of *n* past states will be encoded. This means that the state (as far as the agent is concerned) will only change every *n* characters. However, we will imagine that the agent believes (correctly) that it has *n* opportunities to act before the state changes. An immediate problem is that only the first action after the agent gets a new *n*-gram code can be made with sure knowledge about the immediately preceding state. The agent will have to guess about what has happened until the next code arrives. Another problem is that the agent can only know when it produced a "p" every *n* steps. Given that the effect of the agent's actions and the satisfaction of the agent's goals depend on individual characters, an *n*-gram encoding (where *n* is greater than 1) will only obscure what it is that the agent needs to know. At the same time, we know that a two-gram encoding will be more efficient in representing the stream of characters produced by the Markov chain since the probability of each character depends on the preceding character.

Given this, if our agent must choose between *predicting* events from the character (i.e., one-gram) stance or the two-gram stance, it should certainly choose the former. In contrast, if our agent must choose between *describing* events from the character stance or the two-gram stance, it should choose the latter. This kind of choice is countenanced by Dennett's (1991) original presentation of real patterns in its discussion of an agent's choice of ontology or interpretation scheme.[6] In regard to predicting states of a chess-playing Turing machine realized in Conway's Game of Life, he states:

> Suppose, then, that such an entity occupies the Life plane, playing chess against itself . . . As a first step, one can shift from an ontology of gliders and eaters to an ontology of symbols and machine states, and, adopting this higher design stance toward the configuration, predict its future as a Turing machine . . . Once one has fixed on an interpretation scheme, permitting one to say which configurations of pixels count as which symbols . . . one can use the interpretation scheme to predict . . . the next configuration to emerge. (Dennett 1991, 41)

In this passage, Dennett considers three representations of events. The elementary events comprise the states (on or off) of individual cells. One level up are events that comprise the phases and locations of structures (e.g., gliders or eaters). Two levels up are events that comprise the machine state and tape state of a Turing machine configuration.

Each of the higher-level interpretations considered here departs from the Platonic ideal. Some interpretations (e.g., in terms of Turing machine configurations) are *lossy*—that is, they are not sufficient to recover the exact states of all the cells or exact phases and locations of all the structures. In addition, even if the states of all the cells could be inferred exactly from the Turing machine configurations, this information would still be *implicit* in the higher-level configuration-based representation just as character information is implicit in two-grams. Fortunately for us, these configurations can be used to predict subsequent configurations without extracting the lower-level information. The same is true probabilistically for even higher levels (e.g., the level of beliefs and desires assumed in taking the intentional stance). Whatever predictive value these higher-level representations have for us as agents, the fact

6. I have been speaking about what agents should "choose" to do with respect to representing their environment. Naturally, these "choices" will usually be made by developmental, cultural, or evolutionary processes which the agent cannot fully (or even partially) control. To borrow Dennett's terminology, I have been reasoning from the *design stance*. Machine learning provides rigorous examples of design stance reasoning applied to representation choice. Here, data scientists reason explicitly about about what kinds of data representation will facilitate prediction via formal modeling. Elsewhere (Millhouse 2021), I argue that these practices illustrate an important division of labor between feature selection and model selection/optimization, with the former determining a representation of the data and the latter determining a model for making predictions under that representation.

that later events are predictable from earlier events is a sure sign that we have departed from the Platonic ideal.

5.5 Real Patterns Revisited

As we have just seen, the representations of events that allow an agent to make successful predictions and to take actions to achieve its goals *cannot* be those that it would employ for maximally efficient communication. The optimal approach to representation choice for prediction is beyond the scope of this chapter and will doubtless depend on an agent's goals and mode of interaction with its environment, often called the "task environment" in the context of artificial agents (Russell and Norvig 2010, 40). Nevertheless, we can say some general things about the approach that agents (or agent designers) should take in developing predictively useful representations. First and foremost, they will want to divide their representational efforts into (1) an encoding of events in the environment and (2) a representation of the regularities in the environment under that representation (Millhouse 2021). The latter can then be used to predict the value of different possible actions.

For example, a representation of one's environment as a Markov decision process (or "MDP") will involve a representation of possible states of the environment and a transition function that maps state-action pairs to probabilities over successor states. The agent's goals can be represented in terms of this model, as an assignment of rewards for transitioning from one state to another via a particular action (i.e., a reward function). Taken together, this model allows our agent to employ dynamic programming algorithms to learn the long-run expected rewards of each state (see, e.g., value iteration). With this learned value function for states combined with the reward and transition functions, the agent can directly compute the expected value of taking each action in each state (i.e., the Q-value of each state-action pair). These values can be used to choose between candidate actions rationally.

Representations of events under this division of labor are fundamentally incompatible with maximizing coding efficiency. They are not designed to exploit event regularities for coding efficiency, but to exhibit those regularities clearly so they can be modeled explicitly by the agent for the purpose of guiding its actions to maximize its long-run rewards. Of course, choosing a representation that exhibits relevant regularities does not preclude all efforts at economy. For example, assigning shorter codes to more frequently encountered states and longer codes to less frequently encountered states wouldn't impair our agent's ability to represent state transition probabilities.

More importantly, the fact that an agent refrains from adopting a maximally efficient representation of events does not imply (1) that there is no perspective from which the agent's representation of its environment is efficient or (2) that efficiency (in this sense) does not also rely on patterns in the target system. A number of formal tools have been developed for doing exactly this (see, e.g., Wallace 2005, Rathmanner and Hutter 2011). While the details of these tools are beyond the scope of this chapter, the basic idea underlying them is that a predictive model is a way of encoding information about future states of the world *given earlier states*. More specifically, we can think of a model as one part of a two-part strategy for representing a sequence of future events (where the second part is a description of past events). Given the same sequence of past events, a simpler model will allow us to encode information about future events more efficiently, and for the reasons laid out by Dennett (1991) and others, this efficiency depends on patterns (i.e., regularities relating past to future events). Further, this efficiency serves the practical interests of real-world agents whose ability to represent relevant regularities under memory constraints is enhanced by representing those regularities more compactly.

Given this interpretation of predictive modeling, we can see that the predictive and descriptive stances involve distinct strategies for representing the stance-taker's environment. However, it is only under the former strategy that agents employ a representation of events under which those events exhibit regularities. The reason for this, again, is to ensure that the relationship between past and future events can be explicitly represented by a model and used to guide action. At the same time, it is not clear that agents have any reasons (other than practical reasons) for favoring one strategy over the other. Both strategies depend on real patterns in the target system, with one representing those patterns explicitly and the other representing them implicitly in the code lengths assigned to different events and in the relationship between the events encoded and elementary events. Further, in the limits of optimal prediction and optimal efficiency, neither strategy leaves any regularity unrepresented or unexploited (respectively).

More fundamentally, it is unclear in what sense either strategy gets anything wrong about the world. Yet, the world could not look more different to agents that view their environment through the lens of each stance. For one, the agent's environment is a sequence of random events—none of which predict anything that follows. For the other, the agent's environment is a sequence of nonrandom events that alter the probability of future events and can be predicted by an explicit model to guide goal-directed action.

5.6 Conclusion

In this chapter, I have observed that agents can represent events in their environment in at least one of two ways—with an eye toward successful prediction and with an eye toward descriptive economy. These predictive and descriptive stances not only fail to converge on similar pictures of the world; they recommend representations of events whose relevant properties are all but perfectly opposed. The descriptive stance favors event representations carefully constructed to eliminate any residual regularities relating one event to another. The predictive stance favors event representations carefully constructed to preserve and clearly exhibit a range of agent-relevant regularities. While both approaches are effective means of encoding accurate information about one's environment, the practical value of the latter is dramatically higher *except* where one's goal is itself efficient coding. As a result, it behooves agents to adopt the predictive stance, and it behooves philosophers to acknowledge that it is only by adopting this stance that the familiar world comes into focus.

Still, there is an enormous difference between (1) recognizing that a predictively optimal representation of events is not maximally efficient and (2) building predictively optimal representations. The latter has (for some time) been the work of biological and cultural evolution and (more recently) has been the work of experts in artificial intelligence and machine learning. As I have argued elsewhere (Millhouse 2021), the demands of the task environment will determine how our representations should deviate from optimal efficiency. For present purposes, I have limited myself to cases where pursuing efficient representations directly undermines the ability of agents to learn and act effectively. In most cases, however, these considerations are intertwined with broader questions about representation choice and narrower questions about the demands of particular models and environments. Disentangling these questions and developing a better understanding of how data representation contributes to successful predictive modeling is a project that would benefit machine learning and artificial intelligence *and* give philosophers of science a clearer picture of the relationship between predictive modeling and ontology.

Appendix

In this appendix, I will prove that using a bit to communicate which of two equiprobable events has occurred will maximize the expected change that bit induces in our partner's event probabilities. For present purposes, let the events in question be e and $\neg e$, where p is the probability of e and $1-p$ is the probability of $\neg e$. The update associated with learning that e is $1-p$, since the probability of e is now one. Similarly, the update associated with learning that $\neg e$ is p, since the probability of e is now zero. Let $f(p)$ be the expected value of the update given different values of p, where:

$$f(p)=p(1-p)+(1-p)p. \tag{5.5}$$

Equivalently, simplifying a bit:

$$f(p)=2p-2p^2. \tag{5.6}$$

We know that $f(p)$ is convex since:

$$f''(p)=-4. \tag{5.7}$$

Hence, we know that $f(p)$ is maximized when (and only when) $f'(p)=0$. We can now solve for the value of p that maximizes $f(p)$:

$$f'(p)=2-4p \tag{5.8}$$

$$0=2-4p \tag{5.9}$$

$$0.5=p. \tag{5.10}$$

We find that the expected update associated with learning e (or $\neg e$) is maximized when the probability of e (and $\neg e$) is 0.5.

References

Cover, T. M., and J. A. Thomas. 2006. *Elements of Information Theory.* 2nd ed. Wiley-Interscience.

Dennett, D. C. 1971. "Intentional Systems." *Journal of Philosophy* 68 (4): 87–106.

Dennett, D. C. 1989. *The Intentional Stance.* MIT Press.

Dennett, D. C. 1991. "Real Patterns." *Journal of Philosophy* 88 (1): 27–51.

Dennett, D. C. 2009. "Intentional Systems Theory." In *The Oxford Handbook of Philosophy of Mind,* edited by A. Beckermann, B. P. McLaughlin, and S. Walter. Oxford University Press.

Fano, R. M. 1949. *The Transmission of Information.* Vol. 65. Massachusetts Institute of Technology, Research Laboratory of Electronics.

Huffman, D. A. 1952. "A Method for the Construction of Minimum-Redundancy Codes." *Proceedings of the IRE* 40 (9): 1098–1101.

Millhouse, T. 2021. "Compressibility and the Reality of Patterns." *Philosophy of Science* 88 (1): 22–43.

Plato. 1925a. *Phaedrus.* Edited by H. N. Fowler. Vol. 9. Plato in Twelve Volumes. Harvard University Press; William Heinemann Ltd.

Plato. 1925b. *Statesman.* Edited by H. N. Fowler. Vol. 12. Plato in Twelve Volumes. Harvard University Press; William Heinemann Ltd.

Plato. 1969. *Republic.* Edited by P. Shorey. Vol. 5–6. Plato in Twelve Volumes. Harvard University Press; William Heinemann Ltd.

Rathmanner, S., and M. Hutter. 2011. "A Philosophical Treatise of Universal Induction." *Entropy* 13 (6): 1076–1136.

Russell, S., and P. Norvig. 2010. *Artificial Intelligence: A Modern Approach.* 3rd ed. Pearson.

Shannon, C. E. 1948. "A Mathematical Theory of Communication." *The Bell System Technical Journal* 27 (3): 379–423.

Wallace, C. S. 2005. *Statistical and Inductive Inference by Minimum Message Length.* Springer Science & Business Media.

6 What Emergence Can Possibly Mean

Sean M. Carroll and Achyuth Parola

6.1 Introduction

Everyone agrees that emergence is important, but they don't agree on what the word should mean. The basic idea is that, when a system can be thought of as a composite of many parts, there can be novel properties or behaviors of the composite system that are not readily predictable from considerations of the properties and behaviors of the parts themselves (Gibb et al. 2019; Wilson 2021). Emergence is relevant when there is both a "micro" (more fundamental, comprehensive, lower-level) description and a "macro" (emergent, coarse-grained, higher-level) description of the same system, both of which capture important properties.

Emergence is ubiquitous, and is crucial to how we deal with reality: We are able to model and predict features of the world with dramatically incomplete information about it. We can talk fruitfully about tables and chairs, not to mention people, without anything close to a complete picture of the elementary particles and forces out of which they are apparently made. Properties like "temperature" or "wetness" or "irritability" do not apply to individual particles but are useful emergent features of the macroscopic world. Even more abstract notions like chance and probability may be thought of as emergent phenomena (List and Pivato 2015).

Ambiguities arise when we try to pin down the precise sense in which emergent properties are "novel." There are different conceptions of emergence, which might be relevant to different circumstances, but which also might require very different ontological frameworks. It is traditional to distinguish between the ideas of "weak" emergence, where higher-level properties follow in principle from lower-level ones, and "strong" emergence, where higher-level properties are truly new (O'Connor 2021). Strong emergence is an especially popular concept in some approaches to understanding consciousness, but it is also invoked in discussions of the origin of life and elsewhere (Chalmers 2006; Tononi et al. 2016; Sharma et al. 2023). There is also a distinction between "epistemic" emergence, which refers to an ability to capture features of systems depending on different levels of knowledge about them, and "ontological" emergence, where emergent properties are thought to really exist in a way that is not reducible to lower-level properties.

While these distinctions can be useful, they remain imprecise, and that imprecision opens the door to unresolved conceptual issues. Perhaps the most notable is the relationship between strong or ontological conceptions of emergence and the scope and success of fundamental physics. Granting that there is a lower-level theory that is successful in some domain of applicability, and that the macroscopic system is composed of lower-level parts, it seems strange to accept that success and nevertheless propose that a successful higher-level theory might not be deducible, even in principle, from the lower-level one. Wouldn't that simply be to say that the lower-level theory is wrong, or at least incomplete, and should be appropriately modified to apply even in macroscopic situations? A human brain, for example, contains electrons, quarks, photons, and other particles and fields described very accurately by the rules of quantum field theory (QFT) (Carroll 2022). Given any particular state of those ingredients, the theory makes clear predictions for how that state will evolve over time. To believe that consciousness is strongly emergent would require that the predictions of this QFT, for example for the behavior of electrons, are incorrect when applied to a situation like a human brain (Carroll 2021). That is certainly conceivable, but it seems hard to imagine how the general principles of QFT fail to apply in this situation, or what the actually correct theory would look like. This difficulty makes it important for strong emergentists to be extremely precise about what they mean, especially in relation to fundamental physics.

Beyond that, there is something vague about the novelty component of emergence. Emergent properties are variously described as unexpected, surprising, or impossible to derive, given only the lower-level description (Bedau 1997). But whether or not something is unexpected or novel seems to be a matter of individual judgment, rather than a rigorous designation. It is even hard to be perfectly precise about when something is impossible to derive; maybe it simply hasn't been derived yet. Even formal attempts at describing emergence still appeal to loose notions (Fletcher 2021). This issue of subjective delineations for emergent phenomena is a concern that has been identified before but nonetheless is pervasive in the philosophical discourse on the topic (Taylor 2015).

6.2 Overview

Our aim in this chapter is to clarify some of these issues by clearly delineating different conceptions of emergence in ways that do not rely on personal judgments, and are explicit about the specific relationships between levels. There have, of course, been numerous attempts at classifying different versions of emergence (O'Connor 2021; Wilson 2021). Our strategy has three particular features.

1. We take a physics-inspired approach by focusing not on properties or causal dependence, but rather on states and their dynamics. We take a theory or model to consist of a specified space of states and some (possibly probabilistic) evolution rule, and investigating how lower-level theories can relate to higher-level ones in these terms.
2. We pay explicit attention to the role of mereology, or the decomposition of systems into smaller constituent parts. In discussions of emergence it is typically assumed that macro systems can be thought of as being constituted by smaller pieces. Formally, this amounts to considering the state of a system as being expressible as a Cartesian product of the states of the individual constituents (which might be particles, or lattice sites, or people, or whatever). We attempt to be clear about the role such a decomposition plays in emergence.
3. We replace, insofar as is possible, notions of surprise and novelty with objective standards of the senses in which emergent theories might be described as new, unexpected, or difficult to foresee.

We hope that this approach illuminates what would be required, at the level of states and dynamics, for different conceptions of emergence to pertain.

With these ideas in mind, we propose the following way of classifying varieties of emergence:

- Type-0 (featureless) emergence: A many-to-one map from a micro theory without subsystems to a macro theory, which commutes with time evolution.
- Type-1 (local) emergence: A many-to-one map that commutes with time evolution where both theories describe collections of subsystems, and macro subsystems are made of specific micro subsystems.
  - Type-1a (direct) emergence: Local emergence where the map is algorithmically simple.
  - Type-1b (incompressible) emergence: Local emergence where the map is algorithmically complex.
- Type-2 (nonlocal) emergence: Similar to Type-1, but allowing for macro entities or dynamics that are defined nonlocally from the micro perspective.

- Type-3 (augmented) emergence: A binary relation introducing new ontological features in the macro theory that are simply absent in the micro theory.

We do not put forward any one of our categories as capturing the uniquely correct formulation of emergence; we rather wish to help organize various conceptions that might be relevant to different contexts.

Our categories Type-0, Type-1, and Type-2 all correspond to different versions of what is traditionally labeled as weak emergence, as the higher-level dynamics are completely determined by those at the lower level. The different types are distinguished by how direct and straightforward it is to deduce those higher-level behaviors. Type-3 is close to strong emergence, in that it demands the introduction of new ontological entities at the higher level, which straightforwardly implies that lower-level dynamics is insufficient to predict the higher-level behavior. The usefulness of our conception is that we can be explicit about what kind of theoretical structure is necessary to account for this situation.

How our categories relate to the epistemic/ontological distinction depends on one's thoughts about ontology. Clearly, augmented (Type-3) emergence is an example of ontological emergence. But as Dennett has argued, it can be completely appropriate to categorize the higher-level entities described by Type-1 and Type-2 emergence as "real," in the sense of his "real patterns" (Dennett 1991). The existence of an emergent description implies that it is possible to coarse-grain states of the micro theory and throw away an enormous amount of information about the microstate, and nevertheless retain somewhat accurate predictive power at the macro level. The ability to make such predictions on the basis of so little data is a highly nontrivial fact about the micro theory, and the resulting coarse-grained structures seem well deserving of being labeled "real." One crucial question we do not address is when such patterns exist, or how to quantify them (Crutchfield 1994; Shalizi and Moore 2003; Krakauer et al. 2020; Barnett and Seth 2023).

6.3 Setup

We imagine that there is a (presumably unknown) way of correctly and exactly describing the real world, and label that theory as Ω. It consists of a space of states W with elements $w \in W$. If classical mechanics had turned out to be correct, the space of states would have the structure of phase space, consisting of positions and momenta for each of the components of the world. Since classical mechanics is not correct, and we don't know the complete correct theory of the world, we will assume that there is some space of states, with specifying it further.

One important assumption that we do make is that there is evolution through time, and the relevant evolution law is Markovian—that is, future evolution only depends on the current state (not, for example, on prior history that is not encoded in the current state). This is a feature of all currently popular approaches to fundamental physics, but might seem to be less compelling at higher emergent levels. Maybe some person's behavior is hard to understand without knowing some of their personal history. But that can typically be accounted for by including memories and influences of that history as part of their current state. We will therefore proceed with the Markovian assumption. This means that states evolve with time under some evolution law:

$$w(t+\Delta t) = E_{\Omega}[w(t)]. \tag{6.1}$$

In general the evolution law need not be deterministic, and we should speak about probability distributions rather than specific states. But this distinction won't be relevant for our discussion, so to keep our notation relatively clean we will write as if everything is deterministic.

Now we imagine two theories that seek to capture some aspects of the world:

- Theory α ("micro"), space of states A, individual states a, and evolution rule $a(t+\Delta t) = E_{\alpha}[a(t)]$.
- Theory β ("macro"), space of states B, individual states b, and evolution rule $b(t+\Delta t) = E_{\beta}[b(t)]$.

For each theory, we again assume that the evolution is Markovian: The state at $t+\Delta t$ is determined by the state at t, without any additional knowledge of the prior history of the system.

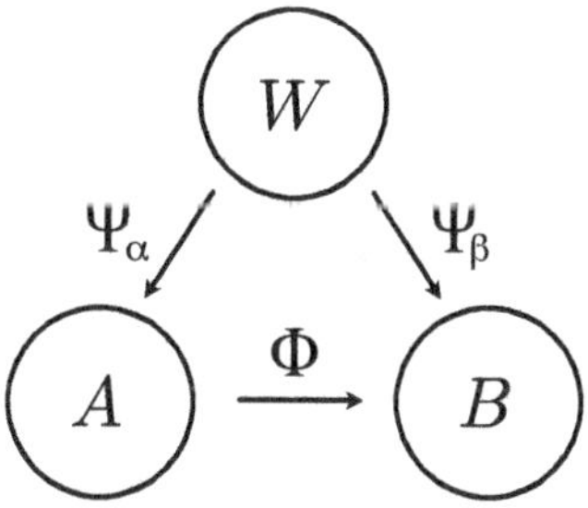

Figure 6.1: Relations between the spaces of states of the world (W), the microscopic theory (A), and the macroscopic theory (B).

The relationship between states of the world and those of these two theories can be illustrated in figure 6.1. In a slight abuse of notation, the arrows represent binary relations between subsets of two spaces of states. Arrows are natural to use because, in many cases, these relations take the form of coarse-graining maps (List 2019), such that one or many states in the lower-level picture get mapped onto a single state of the higher-level picture (as when many states in a collection of atoms are described by the same distribution of density and other variables in a continuum description). But more carefully they should be thought of as binary relations rather than maps. Part of the definition of a map is that elements of the domain are associated with a unique element in the range, but the more general notion of a "relation" simply associates elements of one set with those of another, even if nonuniquely. As we will see in the case of Type-3 emergence below, there could be examples where the ontology of β includes features that are absent in α, so that a single microstate might correspond to multiple possible macrostates. The relations might also have some restricted domain, the "domain of applicability" of the target theory (e.g., fluid mechanics only emerges from kinetic theory when there are sufficiently many particles).

We assume the relations are independent of time. Otherwise emergence would be trivial, as we could always adjust the relations so as to be compatible with time evolution. Note also that we will primarily be concerned about the "emergence relation" $\Phi : A \to B$, but we indicate the separate relations Ψ_α and Ψ_β to emphasize that there might be states in W that are well described by β but not by α (i.e., the domain of applicability of β in W might not be a subset of the domain of applicability of α).

6.4 Type-0 (Featureless) Emergence

In the first case we consider, the space of states A of the micro theory α is not assumed to possess a preferred decomposition into subsystems. The emergence relation Φ is a coarse-graining map, sending sets of microstates in A to single macrostates in B. The criterion that this relation describes emergence is simply compatibility with time evolution, as expressed by the commutativity of the diagram in figure 6.2. Starting from some microstate $a \in A$ at time t, we can either act the emergence map to obtain a state $b \in B$ at t and then evolve forward in time by Δt according to the macro time-evolution operator E_β, or first time-evolve in the micro theory using E_α and then apply the emergence relation Φ. Commutativity, the statement that we end up with the same final macrostate $b(t + \Delta t)$ either way, is the statement that the theory β emerges from α. Then we have:

- **Type-0 (featureless) emergence:** A map Φ (within a specified domain of applicability) from states $a \in A$ in a micro theory α to states $b \in B$ in a macro theory β, such that Φ commutes with the corresponding time-evolution operators E_α and E_β.

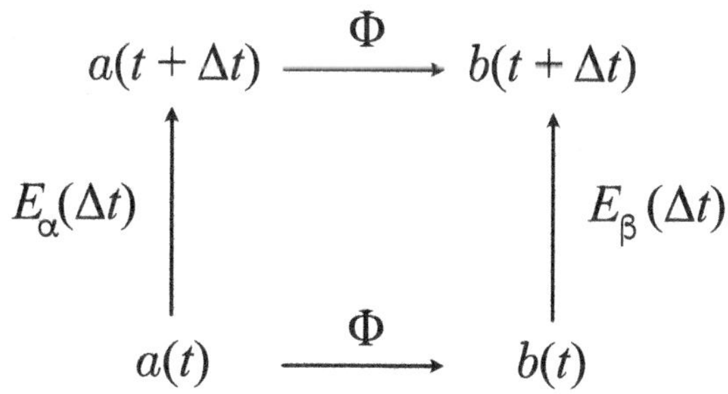

Figure 6.2: The commuting diagram at the heart of emergence, relating time evolution and the emergence map.

We do not insist that every state $a \in A$ in the micro theory have a corresponding macrostate $b \in B$; there may be a restricted domain of applicability $A_E \subseteq A$ where the emergent macro theory is accurate. For that matter, that accuracy may only be approximate rather than exact. It is typical of coarse-grained dynamics that they are only accurate up to some quantifiable level of precision, or with some quantifiable probability.

Type-0 emergence differs somewhat from conventional discussions of emergence, which take for granted a decomposition of the micro theory into subsystems, which then become "parts" of the macroscopic "whole." But it shares the crucial feature of exhibiting efficacious dynamical laws even when microscopic information is discarded (coarse-grained over). This is the essence of a real pattern. It is not obvious that such effective macro dynamics should necessarily or typically exist; throwing away information about microstates generally makes accurate prediction impossible, so that this kind of emergence represents a special situation where very specific kinds of information can be ignored while maintaining macroscopic predictability.

The idea that macro systems are made of parts described by microscopic dynamics is so prevalent that one might wonder whether there are any interesting examples of Type-0 emergence, but a moment's reflection reveals a paradigmatic case: quantum mechanics. Quantum states are represented by vectors $|\Psi\rangle$ in Hilbert space $\mathscr{H}$, a complete normed complex vector space. For a collection of n featureless nonrelativistic particles, this state vector can be represented by an normalized complex-valued function of the particle positions, $\Psi(x_1, x_2, \ldots x_n)$. The phenomenon of entanglement arises because this wave function will generally not be separable into a product of functions for each particle: $\Psi(x_1, x_2, \ldots x_n) \neq \Psi_1(x_1)\Psi_2(x_2) \cdots \Psi_n(x_n)$. Therefore, the quantum system does not naturally decompose into distinct parts with independent dynamics; in general the entire wave function matters.[1]

In certain special circumstances, however, Type-0 emergence occurs: namely, the classical limit. If the wave function happens to be separable or nearly so, and the individual wave functions are relatively localized (and remain so, as will happen for high-mass particles) compared to distances over which the potential-energy function $V(x_1, x_2, \ldots x_n)$ varies, then Ehrenfest's Theorem implies that the expectation values of the positions and momenta, $\langle \hat{x}_i \rangle = \langle \Psi | \hat{x}_i | \Psi \rangle$ and $\langle \hat{p}_i \rangle = \langle \Psi | \hat{p}_i | \Psi \rangle$, approximately obey the classical equations of motion:

$$m_i \frac{d\langle \hat{x}_i \rangle}{dt} \approx \langle \hat{p}_i \rangle, \quad \frac{d\langle \hat{p}_i \rangle}{dt} \approx -\frac{\partial V}{\partial x_i}. \tag{6.2}$$

This is clearly a coarse-graining map: a wave function $\Psi(x)$, specified by a continuum of complex numbers, maps to a discrete set of positions and momenta. Many distinct wave functions will map to the same classical position and momentum. But the wave function doesn't have "parts" in the usual sense, so this is an example of Type-0 emergence. Note the crucial role being played here by the restricted domain of applicability of the emergence map: Classical mechanics only emerges from quantum mechanics under the right circumstances, such that entanglement and interference play no important role in the dynamics.

Other examples include other kinds of limiting behavior, such as the Newtonian limit of relativistic mechanics, when all relative velocities are much less than the speed of light. There, within the domain of applicability, the relevant map is one-to-one rather than many-to-one (relativistic positions and momenta map to nonrelativistic equivalents), so this would be a somewhat trivial example of emergence. There is also the Newtonian-gravity limit of general relativity, which requires both nonrelativistic velocities and a weak gravitational field. One could think of this as a many-to-one map, where the many macroscopically equivalent states of the micro theory (general relativity) are distinguished by negligible but nonzero amounts of gravitational radiation.

1. We may nevertheless decompose Hilbert space into subsystems, but entanglement implies that this will be given by a tensor product structure rather than the simpler Cartesian product. It then becomes an interesting question to ask how best to perform such a decomposition in different physical situations (Carroll and Singh 2021).

Type-0 emergence, based on nothing more than a commutative diagram featuring time evolution and the emergence map, is the basis for the more elaborate Type-1 and Type-2 emergence to be discussed below. The defining feature of emergence is the existence of an effective macroscopic theory, where "effective" means that the dynamics of the macrostates can be described (at least to a good approximation) in terms of those macrostates alone, without reference to the lower-level microtheory.

This kind of commuting relation between two theories, while representative of distinctions in scientific disciplines, is also captured by other notions that relate system descriptions. Bickle's construction of functional, explanatory, and even mechanistic reduction can all be interpreted as various types of mappings through different procedures of coarsegraining, all of which fit the Type-0 constraints (Bickle 2019). Certain notions of computation also depend on commutativity between analytical descriptions and real phenomena (Horsman et al. 2014). Thus, it is worth noting that while relations between scientific theories present the most clear example, Type-0 and following emergence taxonomy will be applicable as broad meta-classifications for a myriad of theory-theory relations.

In terms of the traditional strong/weak dichotomy, Type-0 is a subset of weak emergence. Using Bedau's conception (Bedau 1997), if we knew the micro theory, the microstates, and the emergence map $\Phi : A \to B$, we could put the micro theory on a computer and infer the macro behavior. Whether that behavior would be surprising or not is somewhat a matter of judgment, and we won't draw that distinction quite yet.

6.5 Subsystem Decomposition

For Type-1 and higher versions of emergence, we consider cases where the micro and macro theories describe collections of interacting subsystems, and ask how those subsystems connect to each other under the emergence relation.

First we need to think about what it means to have a subsystem structure. In many traditional examples, what counts as a subsystem is taken as given: One starts with a set of constituents, and collects them together to form a larger system. But since notions of subsystems and locality will be crucial in what follows, it will pay to be specific about what these notions require and imply.

We will model a decomposition into subsystems as a Cartesian product. That is, the space of states of the larger system is simply by an ordered sequence (or tuple) of states of all the subsystems. Formally, for our lower-level theory α we have an injective map from the space of states A into a Cartesian product of I spaces A_i of subsystem states a_i,

$$D_\alpha : A \to \prod_{i=1}^{I} A_i, \tag{6.3}$$

so that individual states are written as I-tuples,

$$D_\alpha : a \to (a_1, a_2, a_3, \ldots a_I). \tag{6.4}$$

In this notation, a_i is the specific state of the ith subsystem, and so on. A similar structure will hold for the higher-level theory β, which is decomposed into $N \leq I$ subsystems:

$$D_\beta : B = \prod_{n=1}^{N} B_n, \qquad b \to (b_1, b_2, b_3, \ldots b_N). \tag{6.5}$$

Cartesian products are not the only way to decompose a large system into subsystems; the notable example is quantum mechanics, where the overall system is described by the tensor product of the subsystems. But other than the Type-0 case of quantum $\to$ classical emergence discussed above, systems of interest are generally decomposed into Cartesian products.

However, there will generally be a very large number of conceivable decompositions of the form (6.3), so we need some way of picking the right one. Our informal notion is that each subsystem should have a kind of identity of its own, and that the dynamics of the system as a whole can be thought of as described by the individual dynamics of the subsystems plus interactions between them. To make this somewhat more explicit, imagine writing the evolution law for each individual subsystem in the microtheory as:

$$a_i(t + \Delta t) = E^{(\text{self})}_{\alpha,i}[a_i(t)] + E^{(\text{int})}_{\alpha,i}[a_i(t), \{a_{\bar{j}(t)}\}]. \tag{6.6}$$

Thus, the evolution of a_i includes a self term for the subsystem under consideration, as well as an interaction term specifying the influence of other subsystems, with the notation $\{a_{\bar{j}}\}$ denoting the set of subsystems that interact directly with a_i. The former depends only on the current subsystem state a_i, while the latter depends also on the state of other micro subsystems.

The expression (6.6) is completely general; it merely reiterates the idea that the evolution of each subsystem depends on the state of that subsystem as well as that of all the other subsystems. To pick out a useful decomposition, we need to implement the idea that the subsystems have individual identities. This amounts to two requirements: (1) We can imagine turning off the interactions with other subsystems, either by "moving them far away" or by setting some coupling constants to zero, and (2) when those interactions are turned off, the data in the subsystem state $a_i(t)$ is sufficient to determine the evolution of that subsystem. When these requirements are met for all the subsystems, we say we have an appropriate subsystem decomposition.[2]

These requirements serve to exclude ways that, in principle, we could decompose a space of states into a Cartesian product, but where the factors fail to represent anything we would recognize as a subsystem. For example, in the classical mechanics of a n particles moving in d spatial dimensions, where the state consists of the positions and momenta of all the particles, formally we could think of the set of all position variables as one subsystem and the set of all momenta as another. But nobody is tempted to do that. Mathematically we can perform such a factorization of phase space, but no particular insight is gained; moreover, knowing the position of a particle but not its momentum (or vice-versa) doesn't allow us to say anything about its future evolution. Under the right conditions, however, it does make sense to treat position/momentum pairs $\{\vec{x}_i, \vec{p}_i\}$ as describing subsystems: to wit, individual particles.

A decomposition into subsystems allows us to think of the space of states as a network, with subsystems as nodes and interactions between subsystems represented by weighted edges. In practice, interactions between subsystems will have different strengths, and some barely interact at all, so that there is a notion of which other subsystems are most relevant to the dynamics. This hierarchy of dynamical relevance can be thought of as a notion of locality, with subsystems that interact noticeably thought of as "near," and less relevant ones as "far," although in higher-level theories those descriptions might not correspond to ordinary physical distance.

In lower-level models based on fundamental physics, spatial locality implies a natural notion of subsystems. Consider for example the Ising model, consisting of spins defined on a lattice with direct interactions only with neighboring spins. Then any particular spin is not directly affected by what is happening on distant lattice sites, only through the intermediaries of neighboring spins. Classical field theories work similarly, where the dynamics of a field at one point in spacetime is directly affected only by the values of various fields (including itself), and their spacetime derivatives, at that same point and no others. The derivatives are the continuum version of "neighboring lattice sites," since they encapsulate infinitesimal changes from point to point.

But our notion of subsystems is more general than locality in space. Consider a set of particles moving through space and interacting via Newtonian

2. In practice we might also require some notion of maximality to our decomposition, so that no subsystem would be better thought of as the union of two smaller subsystems. But the appropriate notion of maximality will generally depend on other considerations (otherwise everything would be divided into elementary particles and fields), so we won't elaborate on that idea here.

gravitation (an inverse-square law).[3] Unlike with a lattice model, now the question of whether subsystems (the particles) are interacting strongly or not depends on the state (in particular, the distance between the particles). There is nevertheless a sensible notion of subsystem individuality for particles, as we can always consider a decoupling regime for one particle where all other particles are far away and the trajectory of the particle under consideration depends only on its own state (and perhaps on a background gravitational field). That stands in contrast with a misguided attempt to think of position and momentum as subsystems; without knowing the momentum of a particle, we can't predict its subsequent position at all.

As a final example, take a network of people interacting over the internet. Individuals will directly interact with some other individuals but not others, but the strength of those interactions is not directly tied to physical distance. A useful decomposition into subsystems, in other words, plays nicely with the dynamics, giving some notion of autonomy to each subsystem under the right conditions.

6.6 Type-1 (Local) Emergence

With that in mind, we define:

- **Type-1 (local) emergence:** Type-0 emergence (i.e., there exists a map $\Phi : A \rightarrow B$ that commutes with time evolution), with the additional condition that subsystems in the macro theory consist of localized collections of subsystems in the micro theory.

"Consist of" here means that the state b_n of each specific subsystem B_n in the emergent theory depends solely on the states $\{a_i^{(j)}\}$ of a specified collection of subsystems of the micro theory $\{A_i^{(j)}\}$, such that each micro subsystem only contributes to the state of one macro subsystem.

Type-1 includes many standard examples of weak emergence, especially in physics. Consider the Ising model mentioned above, or analogous lattice models with nearest-neighbor interactions. As suggested by Kadanoff in early discussions of the renormalization group, we can partition the original lattice into $n \times n$ regions, and associate with each region a "block spin" given by the average of the actual spins. We can then discuss the dynamics of the theory in the long-distance limit by referring only to the block spins and their

3. It is possible to define Newtonian gravity using a potential field obeying a differential equation, as shown by Pierre-Simon Laplace. But the field is not an independent dynamical degree of freedom, as it is completely determined by the distribution of matter (unlike in general relativity). So there is a sense in which the theory is nonlocal.

interactions, in a direct example of emergence (Kadanoff 1966). More generally, effective quantum field theories with an ultraviolet energy cutoff can be thought of as locally emergent theories valid below the cutoff (Burgess 2007).

Another example is provided by center-of-mass motion in classical Newtonian mechanics. In order to predict the motion of the Earth around the Sun, we needn't know the state of every particle constituting the Earth; we need only know the center-of-mass coordinates and momenta of the Earth and other solar system bodies. Indeed, the center-of-mass example was used by Dennett as a paradigmatic instance of a real pattern (Dennett 1991). From our perspective, this reflects the fact that we can make accurate predictions for the trajectories of celestial bodies even after coarse-graining their microstates to just the center-of-mass data, thereby ignoring the vast majority of information needed to characterize the exact microstate. It is crucial that we coarse-grain in the right way; throwing away arbitrary subsets of the micro data would leave us completely unable to make reliable predictions.

In the case of the Ising model or center-of-mass motion, the basic ontology of the macro theory is structurally the same as that of the micro theory. There is nothing in our definition of Type-1 emergence that demands this. Another common example of emergence is the relationship between a microscopic theory of many interacting particles and a macroscopic fluid description. In that case the emergence map takes a small volume of space and calculates appropriate averages over the states of the particles within that volume to determine macroscopic quantities like pressure and temperature, which can be treated as a continuum in the limit of taking the averaging volume to be small (but large enough to contain many particles). A fluid is ontologically different from a set of discrete particles, but this is nevertheless a straightforward example of Type-1 emergence.

One subtlety here is that *which* micro subsystems contribute to the state of a specific macro subsystem may (or may not) depend on the overall state of the micro theory. In the particles-and-fluids case, for example, individual particles (subsystems from the micro perspective) will move from one small averaging volume into others. This is fine, as it doesn't conflict with the idea that the macro subsystem is made out of particular micro subsystems. Similarly, a celebrated example of emergence is Conway's Game of Life, a two-dimensional cellular automaton featuring collective structures such as blinkers and gliders (Izhikevich et al. 2015; Bedau 1997). Since gliders move, changing which set of cells they occupy, the micro systems (lattice sites) that make them up will also change, but that is consistent with our definitions. The Game of Life is also an example of how the macro theory can exhibit complex behavior when the micro rules are relatively simple, although that phenomenon doesn't play any central role in how we have characterized emergence here.

Defined in this way, either Type-0 or Type-1 emergence falls under most traditional notions of weak emergence: We could take the state and evolution rules of the micro theory, put it on a computer, and get behavior that is compatible with the prediction of the macro theory. We nevertheless think the distinction is worth drawing, since what we have called Type-0 is not always recognized as emergence at all, and explicitly highlighting the role of a local decomposition will be important when considering further varieties of emergence.

6.7 Novelty and Compressibility

Despite the straightforwardness of Type-1 emergence, there can be cases falling under this category in which the emergent theory and its properties might nevertheless be plausibly described as "novel" or "surprising" or "unexpected" or even "unpredictable" (Anderson 1972). But these words seem to be about human judgments rather than objective qualities. What if someone fails to be surprised, or claims to have expected the emergent behavior all along? Couldn't any behavior in a locally emergent macro theory be predictable, in principle, by first following the dynamics in the micro theory and then applying the emergence map?

We can try to capture some of this feeling in a more objective way by considering properties of the actual emergence map: in particular, whether it is simple or complex. We are thinking here of algorithmic complexity in the sense of Kolmogorov, Chaitin, and others: the length of the shortest computer program that will produce a desired output (Li, Vitányi, et al. 2008). When there is a compact formula that produces a desired output (such as a string of one billion zeros), the output is said to be simple or compressible; when the shortest program contains an explicit representation of the output and simply prints it (such as a specific billion-digit number with no special properties), the output is complex or incompressible.

In our case, the algorithm in question is the emergence map $\Phi : A \to B$. That map, typically many-to-one, may be expressible in terms of a compact equation, or it might require explicit delineation of which microstates get mapped to which macrostates (or various degrees in between). In the case of center-of-mass motion or particles-to-fluids emergence, for example, it is straightforward to give explicit formulae that construct macrostates from microstates; the mass density is simply the number of particles per volume times their individual masses, the pressure is related to the kinetic energies of the particles, and so on. Whereas in the Game of Life, the definition of a glider cannot be made much more compact than an explicit representation of the appropriate states of the individual cells. This leads to the following subclassifications of local emergence:

- **Type-1a ("direct") emergence:** Type-1 emergence based on a compressible map Φ.
- **Type-1b ("incompressible") emergence:** Type-1 emergence based on an incompressible map Φ.

Admittedly, this distinction is a spectrum rather than a clear-cut binary; some maps may fall in between extreme simplicity and utter incompressibility. But the level of compressibility is at least an objective feature of the map, rather than an expression of our attitudes toward it.[4]

The direct/incompressible distinction is meant to capture some of what is meant by the novelty or unexpectedness of emergence. When the emergence map is direct, we tend to think of the resulting emergent behavior as unsurprising; the existence of a simple map tracks our intuitive feeling for what we should expect from collective behavior. The ontology of fluid mechanics may be different from that of kinetic theory, but the physical behavior of fluids makes sense to us once we learn that fluids are made of molecules. Whereas the need for an explicit, incompressible emergence map tends to go hand-in-hand with emergent behavior being surprising or unexpected. It's not that most collections of micro subsystems would behave in some particular macroscopic way, but that very particular ones do. That kind of behavior is harder to intuitively perceive from the macro point of view.

6.8 Type-2 (Nonlocal) Emergence

The next type of emergence we consider is a different refinement of Type-0, one where locality at the micro level is not manifest at the macro level.

- **Type-2 (nonlocal) emergence:** Type-0 emergence where micro and macro theories have decompositions into local subsystems, but where the macro theory does not respect the notion of locality inherited from the micro theory.

There are basically two ways this can happen. First, the macro theory could involve degrees of freedom that are defined, in principle, in terms of the micro state, but they are not made of local collections of micro entities—they are defined globally, in terms of many dispersed parts. Then the macro entities can't be thought of as simply bundles of nearby subsystems acting in a coherent fashion. Second, the macro entities could interact with each other in ways

4. In general, algorithmic compressibility cannot be computed, but it can be approximated (Jansen 2006).

that would seem nonlocal from the micro point of view. In either case we need to know the states of many (or even all) micro subsystems in order to successfully predict the macro behavior—there is emergence, but not independent emergence of distinct groups of subsystems. We will not introduce separate subcategories for these two possibilities, as in practice they will generally appear together. (The direct/incompressible distinction that we introduced for Type-1 emergence can also be considered in the context of Type-2, but we won't introduce explicit subcategories for this, either.)

This type of emergence generally doesn't happen when the macro theory is not too far removed from the particles and fields of fundamental physics. In such cases, the strict locality of the micro theory tends to be directly carried over to the macro theory, placing limits on the existence of truly nonlocal entities even at the emergent level. In a ferromagnet, for example, there can be nonlocal-seeming *behavior*, when interactions between spins lead to an overall direction of magnetization that is shared throughout the material. This is an example of spontaneous symmetry breaking, where individual solutions to the equations do not respect the full set of symmetries of the equations themselves. But the dynamics remains local, even at the emergent level.

Nonlocal emergence can become relevant, however, in contexts like biology or social sciences. The underlying limit on information propagation imposed by the speed of light is still present, but becomes largely irrelevant, essentially because of the timescales involved. For all practical purposes, information can be shared across distances much faster than the dynamical timescales describing the motions of macroscopic entities, so there is no inconsistency in having emergent entities that are not effectively localized in space. In psychology and cognitive science, for example, our higher-level theory of human beings might contain things like mental states, which cannot be associated with localized groups of neurons or cells, much less with atoms or particles. (They can be associated with appropriate states of *all* the neurons, or all the particles, just not with localized groups of them.) Examples would include Searle's biological naturalism (Searle 2007) and the Integrated Information Theory approach to consciousness (Tononi et al. 2016). Likewise if some sort of social or group dynamics are the higher-level theory, it is natural to include global variables as part of the most parsimonious emergent description. A treaty agreement might play an important role in a causal theory of international relations, but as an emergent entity it can't be identified with a particular collection of micro entities.

By our definition of Type-2 emergence, there can be nonlocal variables in the macro theory, but they are ultimately determined by the state of the micro theory. In that sense they are not truly new. Type-2 does, nevertheless, open the possibility that nonlocal effects feed back onto the micro dynamics in ways that might not have been obvious from a purely micro perspective. It can therefore seem as if the macro theory exhibits behaviors that the micro theory cannot account for, but can ultimately be understood by saying that the micro theory was simply incomplete, rather than incorrect.

Consider the following possible evolution law for a micro subsystem $a_i(t)$:

$$a_l(t+\Delta t) = E^{(\text{self})}_{\alpha,i}[a_l(t)] + E^{(\text{int})}_{\alpha,i}[a_l(t), \{a_{\bar{j}(t)}\}] + \sigma_l[a(t)]E^{(\text{nl})}_{\alpha,i}[a_l(t), \{a_{\bar{j}'(t)}\}]. \tag{6.7}$$

The first two terms are conventional, and are familiar from (6.6): The evolution of a_i includes a self term for the subsystem under consideration, as well as an interaction term specifying the influence of other subsystems, with the notation $\{a_{\bar{j}}\}$ denoting the set of subsystems that interact directly with a_i. The former depends only on the current subsystem state a_i, while the latter depends also on the state of other micro subsystems (with less influence from subsystems that are farther away).

The third term is the new contribution allowed in Type-2 emergence. It includes a new nonlocal interaction factor $E^{(\text{nl})}_{\alpha,i}$ that depends on both the subsystem itself and a (potentially different) set $\{a_{\bar{j}'}\}$ of other subsystems. This novel term is multiplied by a filter function $\sigma_i[a(t)]$ that depends on the global micro state. The filter function goes to zero in situations that are exclusively in the micro domain, such as when a small number of particles are interacting in a physics experiment. But it becomes nonzero in special circumstances that can depend on the global state of the system, perhaps in subtle ways.

In other words, in what we are calling Type-2 emergence, it could indeed be true that electrons behave differently inside a human brain than would be predicted by quantum field theory as we know it. The filter function σ_i might turn on when atoms are arranged brain-wise, and remain zero otherwise. But in laboratory experiments involving relatively small numbers of constituents, including high-energy collisions at particle accelerators, their behavior could remain correctly described by conventional QFT to very high accuracy, essentially using just the first two terms in (6.7). The difference is (or might conceivably be, at any rate) that the dynamics is affected in ways that are undetectable in a variety of conventional experimental conditions. Particle-physics experiments are, quite sensibly, usually performed on small numbers of particles at a time.

To be clear, everything we know about quantum field theory assures us that this doesn't happen: The knowledge we gain about particle interactions from such experiments should allow us to make accurate predictions even when many particles come together in brain-like configurations (Carroll 2021, 2022).

But perhaps what we think we know about QFT is wrong in these situations. There is no principled reason to suspect this from a purely particle-physics perspective, but it is a logical possibility.

But there are other cases where Type-2 emergence is perfectly compatible with physics as we know it. DeDeo (DeDeo 2018) highlights the example of the "jerk"—the derivative of the acceleration with respect to time (i.e., the third derivative of position). This is not part of the fundamental ontology of classical mechanics, which is defined in terms of positions and velocities (or momenta) of the relevant degrees of freedom. Acceleration, the first derivative of velocity, can be calculated in terms of the instantaneous state of the system ($\vec{a} = \vec{F}/m$), but jerk cannot; one would have to follow the system for some bit of time. Nevertheless, we experience the jerk in our everyday lives (e.g., in the motion of a car or an elevator). This is a case of Type-2 emergence, with the lower-level theory given by classical mechanics and the higher-level theory given by human-level experience. One way of thinking about it is that the jerk is not calculable in terms of local and instantaneous quantities of the classical state, but could be calculated if one were to coarse-grain in time as well as in space. However, that is not strictly necessary; information contained in the *global* instantaneous state suffices, according to the rules of classical physics, to determine all of the future evolution of the system. In our setup, the jerk experienced by a macroscopic person supervenes on some not-entirely-local set of micro data (e.g., the state of a car and the driver and the road conditions, as well as the local state of the passenger). In relativistic mechanics the relevant amount of nonlocal information is restricted to the interior of the past light cone of the subsystem in question, but even in nonrelativistic mechanics there will be an effective set of not-too-far-away influences that are relevant to determining the jerk.

Discussions of emergence frequently touch on the possibility of downward causation: higher-level entities exerting causal influence on lower-level entities. Strictly speaking, such a phenomenon is incompatible with either Type-1 or Type-2 emergence as we have defined it; in both cases the dynamics of the micro subsystems are fully determined in their own right. But the possibility of novel macroscopic interactions such as in (6.7) in Type-2 emergence can lead to a kind of counterfeit downward causation. To a macro observer, it might appear as if higher-level features are directly influencing behaviors of the micro systems, even though in principle the micro dynamics are entirely self-contained.

6.9 Type-3 (Augmented) Emergence

All of the varieties of emergence we have thus far considered would traditionally be classified as weak emergence under most definitions thereof. In each case, the higher-level dynamics supervene on the lower-level degrees of freedom. The connection between the levels may be straightforward or it may be subtle, but it is nevertheless ironclad.

Strong emergence is supposed to be different. It posits macroscopic effects that cannot be thought of, even in principle, as simply arising from the collective behavior of smaller parts obeying the microscopic dynamics. The whole is truly different than the sum of its parts.

Strong emergence is most often taken seriously in approaches of consciousness—for example, Chalmers's solution to the hard problem or theories of emergent panpsychism (Chalmers 1995; Goff et al. 2022). In our terminology, these theories attempt to describe a macro model involving consciousness that stands as strongly emergent with respect to a micro model of the world (for the aforementioned theories, this tends to be the traditional physical/biological model of reality and conscious entities). In these theories of consciousness, some form of Type-3 relation between the model that involves consciousness and the physical or biological micro model exists. A new kind of object or entity, independent of and bearing no relation to the decomposition of the micro theory, is put forth as crucial to consciousness.

But the idea of strong emergence is challenging to make sense of. If a higher-level system is a collection of lower-level parts, and we have a dynamical micro theory that accurately describes how those parts behave, then that micro theory will make specific predictions about how the collection will behave. Such predictions are either going to be correct, or incorrect. If they are incorrect, it makes sense to say that the micro theory was just wrong from the start; if they are correct, any macro behaviors should in principle be reducible to the predictions of the micro theory. There seems to be little room for both the micro theory to be correct, and for any new macro behavior we would classify as strongly emergent. In other words, causal closure of the micro theory is incompatible with the kinds of new macro entities required by strong emergence (Kim 2006).

Here we propose a way, within the framework we've been considering, to have a version of strong emergence that avoids this apparent inconsistency. The basic idea is to imagine that there are entities in the ontology of the macro theory that do not supervene on micro subsystems, but whose dynamical relevance disappears in the limit where we consider small numbers of microscopic systems. Then one theory is not derivable from the other, but both represent what a reasonable scientist might develop to describe states and dynamics in the appropriate regime. We refer to this possibility as "augmented" emergence, due to the appearance of these additional macro entities. In this way, the micro theory could be used to make predictions for macro questions, and those predictions would (at least some of the time) turn out to be incorrect, even though the micro theory is perfectly accurate when deployed purely in the micro regime.

To be more specific, we imagine having a micro theory α with a decomposition of states into local subsystems, $A = \prod_i A_i$. This theory is successful as long as we consider situations in which only a few subsystems at a time are evolving and interacting. And there is a macro theory β that is successful as well, but in a domain that (from the point of view of the micro theory) involves a very large number of mutually interacting subsystems (at least for certain states). Because these two domains are distinct, it is possible that there are parts of the ontology of β that have no counterpart in theory α. The space of states of the macro theory is then spanned by both a set of supervenient variables $\{b_n\}$ that are uniquely determined by the micro variables and a new set of independent variables $\{\bar{b}_q\}$ that are not. In that case, the relation Φ associating states in A with those in B need not be a well-defined map, in the sense that single elements of A might correspond to multiple elements of B (ones with different values of the $\{\bar{b}_q\}$). Rather, it can be thought of as a binary relation, connecting microstates to multiple macrostates with different values of the nonlocal variables. We can then define:

- **Type-3 (augmented) emergence:** A macro theory featuring both variables that are constructed from states of the micro theory and new variables that are independent of micro states, but which become dynamically relevant only in certain macroscopic situations.

This corresponds to what is conventionally thought of as strong emergence. The macro theory does not reduce to the micro theory, but the micro theory is not simply wrong; it just doesn't apply in the regime where the macro theory would most naturally be used.

We can think of the dynamics in theories of augmented emergence similarly to the nonlocal Type-2 effects sketched in (6.7). In Type-3, we can express the evolution law for micro subsystems a_i (such as some particular electron) as follows:

$$\begin{aligned} a_i(t+\Delta t) = E^{(\text{self})}_{\alpha,i}[a_i(t)] + E^{(\text{int})}_{\alpha,i}[a_i(t), \{a_{\bar{j}}(t)\}] + \\ \sigma_i[a(t)] E^{(\text{aug})}_{\alpha,i}[a_i(t), \{a_{\bar{j}'}(t)\}, \{\bar{b}_q(t)\}]. \end{aligned} \tag{6.8}$$

As in the Type-2 case, the first two expressions are conventional self and interaction terms. The third term is the new contribution in Type-3 emergence. It includes an interaction factor $E^{(\mathrm{aug})}_{\alpha,i}$ that couples the micro subsystem a_i to the novel macroscopic variables $\{\bar{b}_q\}$, as well as to a possible set of other micro subsystems. The filter function $\sigma_i[a(t)]$ once again goes to zero in situations that are exclusively in the micro domain and becomes nonzero in appropriate circumstances, depending on the global state. In that way, the new augmented variables $\{\bar{b}_q\}$ only become dynamically relevant in those particular circumstances, and could remain invisible to experiments with just a few moving parts.

Once again, from the point of view of physics as it is currently understood, there is no reason at all to suspect something like (6.8) to be true, and many reasons to doubt it, at least when the microscopic variables are fundamental particles and fields. Any full-blown theory along these lines would have to establish that the consequent dynamics are well-defined, that there are no runaway instabilities (ghosts), that gauge invariance and conservation laws are respected (or at least violated only at experimentally permitted levels), and more.

But problems such as consciousness, and perhaps the origin and evolution of life, are difficult, and it is useful to have an explicit framework with which to think about deviations from what we think we know about physics. And something like augmented emergence might be more natural in circumstances where the subsystems of the micro theory are themselves complex, from biological cells to human agents.

6.10 Conclusion

We offer this contribution to an already-considerable literature on emergence in order to help clarify the relationship between philosophical discussions of emergence and a physics-oriented view of states and their dynamics. We take the basic feature of emergence to be the existence of a particular kind of real pattern—a coarse-grained macroscopic theory that is able to make accurate predictions on the basis of much less data than is contained in the microscopic description. By carefully considering the decomposition of systems into subsystems with certain types of interactions, and examining different relationships between such decompositions at the micro and macro level, we are able to classify varieties of emergence explicitly and without leaning on subjective characterizations.

6.11 Acknowledgments

We are very grateful to a number of generous readers who provided invaluable suggestions on the manuscript, including David Chalmers, Simon DeDeo, David Kinney, David Krakauer, and Elanor Taylor. We thank Tyler Milhouse for his encouragement and patience. And of course we offer thanks and appreciation to Dan Dennett, for inspiration and influence both evident and subtle.

References

Anderson, P. W. 1972. "More Is Different: Broken Symmetry and the Nature of the Hierarchical Structure of Science." *Science* 177 (4047): 393–396.

Barnett, L., and A. K. Seth. 2023. "Dynamical Independence: Discovering Emergent Macroscopic Processes in Complex Dynamical Systems." *Physical Review E* 108 (1): 014304. https://doi.org/10.1103/PhysRevE.108.014304.

Bedau, M. A. 1997. "Weak Emergence." *Philosophical Perspectives* 11:375–399. https://doi.org/10.1111/0029-4624.31.s11.17.

Bickle, J. 2019. "Reduction." In *The Routledge Handbook of Emergence,* edited by S. Gibb, R. F. Hendry, and T. Lancaster, 65–76. Taylor & Francis. ISBN: 9781317381501.

Burgess, C. P. 2007. "An Introduction to Effective Field Theory." *Annual Review of Nuclear and Particle Science* 57 (1): 329–362.

Carroll, S. M. 2021. "Consciousness and the Laws of Physics." *Journal of Consciousness Studies* 28 (9–10): 16–31.

Carroll, S. M. 2022. "The Quantum Field Theory on Which the Everyday World Supervenes." In *Levels of Reality in Science and Philosophy: Re-Examining the Multi-Level Structure of Reality,* edited by S. Ioannidis, G. Vishne, M. Hemmo, and O. Shenker, 27–46. Springer.

Carroll, S. M., and Ashmeet Singh. 2021. "Quantum Mereology: Factorizing Hilbert Space into Subsystems with Quasiclassical Dynamics." *Physical Review A* 103 (2): 022213. https://doi.org/10.1103/PhysRevA.103.022213. arXiv: 2005.12938 [quant-ph].

Chalmers, D. 1995. "Facing Up to the Problem of Consciousness." *Journal of Consciousness Studies* 2 (3): 200–219.

Chalmers, D. 2006. "Strong and Weak Emergence." In *The Re-Emergence of Emergence: The Emergentist Hypothesis from Science to Religion,* edited by P. Clayton and P. Davies. Oxford University Press.

Crutchfield, J. P. 1994. "The Calculi of Emergence: Computation, Dynamics and Induction." *Physica D: Nonlinear Phenomena* 75 (1): 11–54. ISSN: 0167-2789. https://doi.org/https://doi.org/10.1016/0167-2789(94)90273-9.

DeDeo, S. 2018. "Origin Gaps and the Eternal Sunshine of the Second-Order Pendulum." In *Wandering Towards a Goal,* edited by A. Aguirre, B. Foster, and Z. Merali, 41–

61. Springer International Publishing. ISBN: 9783319757261. https://doi.org/10.1007/978-3-319-75726-1_5.

Dennett, D. C. 1991. "Real Patterns." *Journal of Philosophy* 88 (1): 27–51.

Fletcher, S. C. 2021. "Similarity Structure and Diachronic Emergence." *Synthese* 198:8873–8900. https://doi.org/10.1007/s11229-020-02605-3.

Gibb, S., R. F. Hendry, and T. Lancaster. 2019. *The Routledge Handbook of Emergence.* Routledge Handbooks in Philosophy. Taylor & Francis. ISBN: 9781317381501. https://books.google.com/books?id=0Tz3DwAAQBAJ.

Goff, P., W. Seager, and S. Allen-Hermanson. 2022. "Panpsychism." In *The Stanford Encyclopedia of Philosophy,* Summer 2022, edited by Edward N. Zalta. Metaphysics Research Lab, Stanford University.

Horsman, D., S. Stepney, Rob C. Wagner, and V. Kendon. 2014. "When Does a Physical System Compute?" *Proceedings of the Royal Society A: Mathematical, Physical and Engineering Sciences* 470 (2169): 20140182. https://doi.org/10.1098/rspa.2014.0182.

Izhikevich, E. M., J. H. Conway, and A. Seth. 2015. "Game of Life." *Scholarpedia* 10 (6): 1816.

Jansen, T. 2006. "Introduction to the Theory of Complexity and Approximation Algorithms." *Lectures on Proof Verification and Approximation Algorithms,* 5–28.

Kadanoff, L. P. 1966. "Scaling Laws for Ising Models Near T_c." *Physics* 2:263–272.

Kim, J. 2006. "Emergence: Core Ideas and Issues." *Synthese* 151 (3): 547–559. https://doi.org/10.1007/s11229-006-9025-0.

Krakauer, D., N. Bertschinger, E. Olbrich, J. C. Flack, and N. Ay. 2020. "The Information Theory of Individuality." *Theory in Biosciences* 139:209–223.

Li, M., P. Vitányi, et al. 2008. *An Introduction to Kolmogorov Complexity and Its Applications.* Vol. 3. Springer.

List, C. 2019. "Levels: Descriptive, Explanatory, and Ontological." *Noûs* 53 (4): 852–883.

List, C., and M. Pivato. 2015. "Emergent Chance." *Philosophical Review* 124 (1): 119–152. https://doi.org/10.1215/00318108-2812670.

O'Connor, T. 2021. "Emergent Properties." In *The Stanford Encyclopedia of Philosophy,* Winter 2021, edited by Edward N. Zalta. Metaphysics Research Lab, Stanford University.

Searle, J. 2007. "Biological Naturalism." In *The Blackwell Companion to Consciousness,* edited by M. Velmans and S. Schneider, 327–336. Wiley-Blackwell.

Shalizi, C. R., and C. Moore. 2003. "What Is a Macrostate? Subjective Observations and Objective Dynamics." *arXiv preprint cond-mat/0303625.*

Sharma, A., D. Czégel, M. Lachmann, et al. 2023. "Assembly Theory Explains and Quantifies Selection and Evolution." *Nature* 622:321–328. https://doi.org/10.1038/s41586-023-06600-9.

Taylor, E. 2015. “Collapsing Emergence.” *The Philosophical Quarterly* 65 (261): 732–753. ISSN: 0031-8094. https://doi.org/10.1093/pq/pqv039.

Tononi, G., M. Boly, M. Massimini, and C. Koch. 2016. “Integrated Information Theory: From Consciousness to Its Physical Substrate.” *Nature Reviews Neuroscience* 17 (7): 450–461. https://doi.org/10.1038/nrn.2016.44.

Wilson, J. M. 2021. *Metaphysical Emergence.* Oxford University Press.

7 Is Any Pattern Nonreal?

James W. McAllister

7.1 Introduction

The concept of pattern has been part of the toolbox of philosophy of cognition and science for over sixty-five years. Its value stems from its hybrid nature: A pattern is neither entirely of the world nor entirely of the knower. First, a pattern inheres in concrete experience, yet simultaneously represents an abstraction from it. Second, while it is an objective feature of a data string, a pattern acquires significance only once an observer individuates it. As a consequence, writers have turned to the concept of pattern as a resource to explain how cognition and science constructed stable, broad structures in the world from strings of observations.

One of the first was Norwood R. Hanson. In *Patterns of Discovery*, Hanson argued that observation in science was theory laden, meaning that what scientists observed was influenced by their theories and conceptual frameworks. As Hanson put it, "Physical theories provide patterns within which data appear intelligible. . . A theory is not pieced together from observed phenomena; it is rather what makes it possible to observe phenomena as being of a certain sort, and as related to other phenomena" (Hanson 1958, 90).

Building on this idea, James Bogen and James Woodward (1988) reexamined the relation between theory and observation. They argued that empirical data, due to their idiosyncratic nature, could neither effectively test scientific theories nor serve as their explananda. Instead, they introduced an intermediary based on the notion of pattern. Patterns in empirical data, they suggested, constituted evidence of what they termed "phenomena"—fundamental and stable modes in which reality was articulated. Phenomena, in turn, were the explananda of scientific theories.

Empirical data exhibit many patterns. The world, however, in Bogen and Woodward's view, comprised a relatively small number of phenomena. This raised a crucial question: How we could tell whether a particular pattern was evidence of a phenomenon? I criticized Bogen and Woodward's proposal for, as I saw it, failing to provide a coherent answer to this question (McAllister 1997).

Daniel C. Dennett (1991) independently proposed a similar role for patterns. His "real patterns" hypothesis suggested that cognition consisted largely in detecting patterns in experience. Among these patterns, some—the "real" ones—corresponded to structures in the world and constituted evidence of them. Other patterns were spurious and did not reflect anything in the world. Dennett recognized the need to distinguish real and spurious patterns, proposing projectability and economy of description as criteria.

James Ladyman and Don Ross (2007) adopted Dennett's real patterns hypothesis within a structural realist framework. Structural realism holds that what is real is not individual entities, but rather structures. These structures are also what scientific theories describe. Patterns in empirical data, Ladyman and Ross argued, were evidence of these fundamental structures. Like Dennett, they acknowledged the need to differentiate real patterns from spurious ones: They refined Dennett's criteria in order to close what they saw as loopholes.

In this contribution to the debate, I critically examine the real patterns hypothesis, focusing on the distinction that it draws between patterns in empirical data that correspond to structures in the world and patterns that do not. I argue that there is no means available to partition the infinite number of patterns that we can discern in empirical data into these two classes. In particular, the criteria that Dennett, Ladyman, and Ross have proposed are insufficient. I argue that, as a consequence, we are obliged to regard all patterns in data as having an equally strong claim to correspond to structures in the world. In short, if any pattern is real, then all patterns must be.

The chapter proceeds as follows. I review the epistemology and methodology of the approach that construes patterns in empirical data as evidence of structures in the world in sections 7.2 and 7.3. One conclusion is that there are infinitely many ways to decompose a dataset into pattern and noise. How should we respond to this multiplicity? In section 4, I explore the hypothesis that only some patterns in data constitute evidence of structures in the world, and I explain why it must offer criteria for distinguishing real and bogus patterns. In section 7.5, I reconstruct the criteria proposed by Dennett, Ladyman, and Ross, and critique them in section 7.6. In section 7.7, I turn to the alternative option: Accepting that all patterns in data constitute evidence of structures in the world. I present positive reasons for this view. In section 7.8, lastly, I summarize and discuss the implications for the real patterns hypothesis.

7.2 Epistemology of Patterns

Empiricism asserts that all our evidence about the world comes in the form of empirical data. The world as a whole generates each measurement outcome, every data point, and thus the entirety of any empirical dataset. In most cases, of course, scientists investigate not the world as a whole but individual structures within it. Since these structures are causal components of the world, we assume that each one gives rise to a component of the empirical datasets that the world generates. Evidence of individual structures in the world consists, therefore, of patterns exhibited in empirical datasets.

More precisely, we assume two reciprocal relations between structures in the world and patterns in empirical data. The first is causal-explanatory: Structures in the world cause, and therefore also explain, patterns in data. The second is evidential: Patterns in data are evidence of structures in the world. The causal-explanatory relation ensures the evidential significance of patterns: If patterns in data were not caused by structures, they would not convey information about them. Conversely, the evidential relation places a condition on the knowability of structures in the world: If a structure caused no pattern in data, we would have no epistemic access to it.

This means that our individuation of structures in the world depends on patterns in empirical data. The only way to ascertain the presence of a particular structure in the world is by finding a pattern in data that we label as its effect. Similarly, any distinction that we make among structures in the world must be reflected in patterns in data if it is not to be otiose. If we distinguish between two structures in the world, S_1 and S_2, then we must be able to indicate a respect in which a pattern in data that corresponds to S_1 differs from a pattern that corresponds to S_2. This requirement holds even when distinctions between structures are first made on theoretical grounds: Even theoretical posits must, sooner or later, demonstrate empirical significance.

Two examples from scientific practice illustrate the relation between structures in the world and patterns in data. In the first example, a distinction between structures reflects a difference between patterns; in the second, the inability to find a difference between certain patterns nullifies a proposed distinction between structures.

Chaos theorists distinguish two classes of phenomenon in the world: Ordered and chaotic. Each of these manifests itself in a recognizable class of pattern in data: Ordered phenomena are exhibited in periodic patterns, which appear as closed loops in state space, whereas chaotic phenomena are exhibited in nonperiodic patterns, which appear as unbounded phase space trajectories (Motter and Campbell 2013). This correspondence links properties of phenomena—ordered and chaotic behaviors—to which we have no direct epistemic access, to properties of patterns—periodicity and nonperiodicity—that are detectable in empirical data. We think that chaotic behavior in a phenomenon causes nonperiodicity in a pattern, and, reciprocally, that nonperiodicity in a pattern is evidence of the chaotic nature of a phenomenon; and the same for ordered behavior and periodicity. A difference among patterns in data empirically grounds a distinction between structures in the world.

Conversely, scientists regard an absence of difference in the former domain as invalidating distinctions drawn in the latter. For example, nineteenth-century physicists were led by theoretical considerations to hypothesize two distinct phenomena of electromagnetic induction, an effect in which an electric current arose in a conductor through interaction with a magnet. These differed depending on whether the conductor or the magnet was at rest or in motion with respect to the reference frame of the electromagnetic ether. The explanations of the rise of the current in the two cases invoked different physical mechanisms, one involving an electromotive force, the other an electric field.

As Albert Einstein noted in 1905, however, the difference between the two purported phenomena did not appear in the empirical evidence. The evidence took the form of patterns in empirical data consisting of galvanometer readings plotted against the velocities of the conductor and magnet. The patterns caused by a moving conductor were indistinguishable from those caused by a moving magnet. Einstein concluded that electromagnetic induction was a single phenomenon dependent solely on the relative motion of conductor and magnet, rendering the concept of ether irrelevant (Hon and Goldstein 2005, 479–491). The lack of a difference between patterns in data that corresponded to the two states of motion relative to the ether made the distinction between the two purported phenomena untenable, despite initial theoretical reasons for it.

These examples reinforce the principle that differences in the domain of patterns in empirical data are our only guide to differences in the domain of structures in the world.

7.3 Methodology of Patterns

Now for methodological aspects of this approach. A pattern exhibited in an empirical dataset is, according to this view, an additive component of it. If a dataset $E(x)$ is a string of digits of length L, then a pattern $A(x)$ within it is a string of length L such that the sum of $A(x)$ and of a second term, $R_A(x)$, also of length L, yields $E(x)$.

Empirical datasets in science typically have high algorithmic complexity, being the outcome of an indefinitely large number of causal factors. In contrast,

$A(x)$ generally has much lower algorithmic complexity—a pattern that we pick out from an empirical dataset typically shows much more regularity than the original data. This means that $A(x)$ can be generated by a formula much shorter than L. The remainder, $R_A(x)$, lastly, has algorithmic complexity approximately equal to that of $E(x)$: If a regular term is subtracted from an irregular term, what is left must be irregular.

The process by which scientists mine empirical data for evidence about structures in the world therefore consists of expressing an empirical dataset, $E(x)$, as the sum of two components: $A(x)$, a term that the scientists propose as evidence of a particular structure in the world, and $R_A(x)$, a remainder. This amounts to decomposing a dataset into pattern and noise. The noise term, $R_A(x)$, as defined here, is nothing other than the mathematical remainder left when $A(x)$ is subtracted from $E(x)$. Whereas some may think of $R_A(x)$ as an error term, there is nothing at this stage to suggest that it is the product of any observational or other error or corresponds to any misdescription of the world: It is simply a component of the empirical dataset not currently of interest.

Mathematically, any empirical dataset of length L can be decomposed into any one of all patterns of length L and the associated noise term. We may express $E(x)$ as $A(x) + R_A(x)$, as $B(x) + R_B(x)$, as $C(x) + R_C(x)$, and so on without end. The patterns thus picked out, $A(x)$, $B(x)$, $C(x)$, and so on, differ in mathematical form and algorithmic complexity; the noise terms with which they are exhibited in the dataset, $R_A(x)$, $R_B(x)$, $R_C(x)$, and so on, vary in magnitude.

Scientists routinely exploit the fact that a given empirical dataset can be decomposed into multiple pattern–noise pairs. They associate each of the patterns in this multiplicity with a different causal structure in the world, which they take to explain the pattern. This underlines two things. First, a typical empirical dataset contains information about many more causal structures in the world than just one. Second, patterns in data that scientists interpret as evidence of structures differ widely in mathematical form, in algorithmic complexity, and in the noise levels with which they are exhibited (Petersen 2023).

Two examples illustrate this multiplicity. The first pertains to spatial variations in sea surface topography. We measure sea surface heights relative to a reference frame based on the lithosphere. A global dataset of sea surface heights at a particular time has high algorithmic complexity. Oceanographers identify several different patterns in this dataset, which they take as evidence of distinct causal factors acting on sea water. One pattern consists of an ellipsoid: Mean sea level at the equator is around twenty kilometers higher than at the poles. Oceanographers attribute this pattern to the earth's rotation. A second pattern consists of "hills" and "valleys" on large geographical scales: For example, mean sea level around Iceland is some 150 meters higher than around Sri Lanka. Oceanographers attribute this pattern to variations in the earth's gravitational field due to inhomogeneous mass distributions in the crust and mantle. Further patterns consist of complex variations on regional and local scales: For example, mean sea level at the Pacific end of the Panama Canal is around 0.2 meters higher than at the Atlantic end. Oceanographers attribute these patterns to a variety of causal factors, including tides, currents, wind forcing, variations in atmospheric pressure, and variations in water density caused by temperature and salinity differences (Pugh and Woodworth 2014, 223–251).

These patterns correspond to alternative ways of accounting for the variance of the sea surface height dataset by decomposing it into a pattern and an associated noise term: They are superposed on one another in the dataset. They illustrate a trade-off, often seen in science, between a pattern's algorithmic complexity and the magnitude of the noise term with which it is exhibited in the data. For example, the pattern consisting of the ellipsoid has low algorithmic complexity and is exhibited in the global dataset with a high noise level: That is, there are relatively large discrepancies between the ellipsoid and actual sea surface heights around the globe. On the other hand, the patterns on the smaller spatial scales have higher algorithmic complexity than the ellipsoid and are exhibited in the dataset with lower noise levels: They track the measurement data more closely.

The second example concerns temporal variations in photosynthetic activity in evergreen conifers. Botanists have identified multiple patterns on different temporal scales in measurements of the photochemical reflectance index, an indicator of photosynthetic activity, attributing each of them to a distinct physiological process. One is a diurnal pattern: Botanists have explained this as the result of interconversion of xanthophyll pigments, which allows the tree to balance photosynthesis and photoprotection under varying light conditions. A second pattern has a period of a few days: Botanists have attributed this to changes in leaf albedo caused by fluctuating water retention in warm and cold weather spells. A third pattern is seasonal: Botanists have explained this as the effect of changing ratios of chlorophyll and carotenoid pigments in the leaves (Wong and Gamon 2015).

Again, these patterns represent different ways of analyzing an empirical dataset into pattern and noise. Each pattern is taken to correspond to a different causal process in the world, each has a different mathematical form and algorithmic complexity, and each is exhibited in the dataset with a different noise level. There are many similar examples in all branches of science (McAllister

2010, 807–808; 2011, 77–79): They illustrate the wide variety of patterns in empirical data that scientists have successfully taken as evidence of structures in the world.

7.4 Some Patterns or All?

As we have noted, there are infinitely many ways to decompose a dataset into pattern and noise. This creates a puzzle for anyone who believes that evidence of structures in the world comes in the form of patterns in empirical data. How many of these patterns are evidentially significant, and which patterns are they?

There are two possible answers: Either only some patterns in an empirical dataset correspond to structures in the world, or all do. Most writers have favored the first view. I shall therefore start by exploring this option, which I shall call "bipartition thesis"; I shall return to the second option in section 7.7.

The appeal of the bipartition thesis is clear. It accords with the view that the world is composed of a univocal, determinate, and perhaps relatively small set of structures that clearly shine through in empirical data. It says that some patterns that empirical data exhibit are caused by these structures in the world, and thereby constitute evidence of them. Other patterns have no evidential significance. It must be that the latter are not caused by anything in the world: If they were, then they would bear traces of their causes and thus constitute evidence of them. They must thus arise spontaneously or randomly in the data.

The bipartition thesis is a special case of a familiar class of claims in science: Claims that a specific set of structures in the world causes some patterns that we find in empirical data but not others. We encountered other cases in section 7.2. Chaos theorists maintain that particular structures in the world—chaotic phenomena—cause some patterns in empirical data but not others. Similarly, some nineteenth-century physicists believed that the phenomenon of electromagnetic induction in which the magnet was at rest relative to the ether caused some patterns in empirical data but not others. The bipartition thesis differs from these claims in scope: It ranges over the totality of causal structures in the world rather than just a particular kind. It does not differ from these more restricted claims in form, however.

As a consequence, the bipartition thesis faces the same challenge as these more restricted claims: It must explain how patterns in empirical data of the two sorts differ. Our earlier discussion of the epistemology of patterns concluded that differences among patterns in data were our only basis for positing distinctions between structures in the world, and any distinction in the latter domain had to be reflected in a difference in the former. If we maintain that chaotic phenomena cause some patterns in data but not others, for example, then we need to identify some difference between patterns caused by those structures and other patterns. Chaos theorists meet this challenge by saying that patterns in data caused by chaotic phenomena are nonperiodic, whereas other patterns are periodic. Conversely, Einstein observed that the absence of differences between patterns in data undermined the distinction between phenomena of electromagnetic induction in which the magnet and the conductor was at rest relative to the ether.

Proponents of the bipartition thesis must similarly demonstrate that the difference between patterns that are and are not evidence of structures in the world is empirically detectable. Failing to identify a difference between patterns of the two sorts would lead to skepticism about causal structures. If nothing distinguished patterns in empirical data that were caused by the world from patterns that were not, what need would there be to posit structures in the world at all? Empirical data would show patterns of the same sorts even if no structures existed to cause them. The bipartition thesis would then undermine itself: An account devised to explain how we could draw inferences about causal structures in the world from patterns in empirical data would end up casting doubt on the existence of structures themselves.

7.5 Candidate Criteria for Real Patternhood

How might we identify a difference between patterns that are evidence of structures in the world and patterns that are not? At the outset, we must acknowledge two restrictions. First, we may not rely on an assumption, expectation, belief, or purported knowledge that the world contains a particular structure. In this metaphysical project, we are investigating which patterns correspond to structures in the world before we know which the latter are; we cannot, therefore, presuppose particular structures in the world. We cannot respond to Einstein that the difference between patterns in electromagnetic induction data is that some patterns result from the motion of the conductor and others from the motion of the magnet relative to the ether: That would presuppose the very distinction that Einstein brought into question. In distinguishing between real and other patterns, similarly, we may not avail ourselves of any conclusion that assumes such a distinction.

Second, a criterion to distinguish between patterns that are and are not evidence of structures in the world must not depend on decisions, choices or stipulations made by us knowers, such as might follow from our cognitive interests or pragmatic preferences. The demarcation must depend solely on how the world is, not on our wishes or conventions. For example, we may not declare that only patterns exhibited in datasets with a noise level below 10% are evidence of structures in the world: A conventional threshold cannot serve as a guide to ontology. A preference that would be legitimate when choosing a pattern for some attractive or advantageous characteristic in order to work on it would not be admissible in the task of demarcating patterns into classes that had and lacked physical significance.

These two restrictions mean that any difference between patterns that have and patterns that lack evidential significance must consist of features that are both objective and detectable in empirical data. Still, these restrictions allow several possible accounts of the difference.

Dennett proposed two criteria for distinguishing "real" patterns in empirical data, which constituted evidence of structures in the world, from "bogus" patterns, which did not (Dennett 1991, 29). On Dennett's first criterion, a pattern in a given dataset was real if it provided a more economical encoding of the data than a bit map, or bit-by-bit transcription of the dataset: "A pattern exists in some data—is real—if there is a description of the data that is more efficient than the bit map, whether or not anyone can concoct it" (Dennett 1991, 34; Millhouse 2021). This wording implied that pattern reality coincided with existence, but the rest of the article made it clear that Dennett thought that bogus patterns existed too.

On Dennett's second criterion, a pattern in data was real if it was projectable, meaning that it could be used to make better-than-chance predictions about future occurrences. As Dennett pointed out, many distinct patterns exhibited in the same dataset with different noise levels could all lead to good predictions, implying that many patterns were real or that real patternhood was relative to an observer's tolerance for noise (Dennett 1991, 35–36).

Structural realists, too, have assumed that only some patterns in empirical data were real or corresponded to structures in the world. As Ladyman and Ross put it, "for a pattern to be real" is "for the object of a scientific theory or other description to be deemed an aspect of objective reality" (Ladyman and Ross 2007, 36).

Structural realists had good reason to take the view that the number of real patterns was limited. Structural realism arose as an alternative to traditional scientific realism. It lowered the bar for a scientific theory to achieve truth approximation: Instead of requiring a theory to describe correctly the entities that the world contained and their properties, structural realists demanded success only in capturing fundamental causal structures (Worrall 1989, 117–121). This meant that theories, to demonstrate truth approximation, no longer needed to account for all details of empirical data, as long as they accounted for significant patterns in them. The smaller the number of fundamental structures and significant patterns, the clearer the difference between structural realism and traditional scientific realism.

Ladyman and Ross agreed with Dennett that, for a pattern to be real, it should support predictions about unobserved occurrences. However, they wished to avoid the two outcomes that Dennett seemed to allow: An explosion in the number of real patterns and the relativity of real patternhood to an observer's noise tolerance (Ladyman and Ross 2007, 206). Ladyman and Ross thought that a conjunction of three criteria would be sufficient to block these outcomes. On their definition, a pattern is real just if:

> (i) it is projectible; and
>
> (ii) it has a model that carries information about at least one pattern P in an encoding that has logical depth less than the bit-map encoding of P, and where P is not projectible by a physically possible device computing information about another real pattern of lower logical depth than [the candidate pattern]. (Ladyman and Ross 2007, 233)

Condition (i) and the first part of condition (ii) incorporated Dennett's criteria of predictive power and economical encoding. The second part of condition (ii) prescribed that no other pattern should offer a more economical encoding, thereby linking real patternhood to predictive indispensability (Suñé and Martínez 2021).

In the next section, I shall examine the two key ideas underlying the approach of both Dennett and the structural realists—projectability and economy of description—as criteria for real patternhood. Before doing so, I will briefly outline two objections to the second part of Ladyman and Ross's condition (ii). I believe that it violates both the preliminary restrictions discussed earlier.

My first objection concerns the reference to a "physically possible device" in condition (ii). This reflects Ladyman and Ross's belief that projectability should be assessed from the perspective of our best current science: As they explained, "A pattern, to be real, must be projectible from a perspective that physics tells us could be physically occupied" (Ladyman and Ross 2007, 236). This reference breaches the first of my two preliminary restrictions, namely that we may not presuppose an assumption, expectation, belief, or purported knowledge that the world contains a particular structure in demarcating real patterns from bogus ones. If we are to demarcate patterns that are evidence of structures in the world from patterns that are not, we cannot help ourselves to conclusions of current science about which structures the world contains. Ladyman and Ross's appeal to current physics does just that.

Second, the effect of the second part of condition (ii) is that no more than one pattern in any given dataset counts as real. If two patterns, $A_1(x)$ and $A_2(x)$, encode the same dataset, but $A_2(x)$ is of lower logical depth or is more economical, then $A_1(x)$ is disqualified as real pattern. Ladyman and Ross offer no reason for thinking that only one pattern in a given empirical dataset is caused by or corresponds to a structure in the world, however. In the absence of such a reason, theirs is a purely pragmatic choice—it makes it easier to pick out a small set of patterns to designate as real. This violates the second of my two preliminary restrictions, namely that we may not allow a demarcation of real from bogus patterns to depend on decisions, choices, or stipulations made by us knowers.

7.6 Projectability and Economy of Description

We now turn to the two main ideas proposed by Dennett and structural realists to differentiate between real patterns from bogus ones: Projectability and economy of description.

The first proposal is that a pattern in empirical data is real, or is evidence of a structure in the world, just if it is projectable. A pattern is projectable if it is exhibited in a dataset broader than the one in which it is first identified, thereby allowing successful predictions. As Ladyman and Ross explained, "We want to associate the distinction between projectible and non-projectible patterns with the distinction between what's real and what isn't" (Ladyman and Ross 2007, 228).

This proposal is based on the intuition that a pattern that leads to a successful prediction has a stronger claim to reality than a pattern that does not. While scientists can easily fit multiple patterns to existing empirical data, we might reason, the stricter test comes when they check these patterns against future data. Patterns that persist in the new data seem more likely to correspond to structures in the world. In Ladyman's view, projectability "provides a criterion of ontological commitment: Real patterns are those that indispensibly figure in projectible generalizations that allow us to predict and explain the behavior of the world" (Ladyman 2017, 157).

But what does saying that a pattern is or is not projectable amount to? Consider a pattern observed in an initial dataset, *A*. This pattern is projectable if it is replicated in a subsequent dataset, *B*. Patterns, however, are exhibited in empirical datasets with certain noise levels, as we know: A more precise definition of projectability needs to reflect this fact. Suppose two patterns, $P(x)$ and $Q(x)$, are exhibited in *A* with a noise level of $n\%$. $P(x)$ is projectable if it is subsequently exhibited with a noise level no higher than $n\%$ also in *B*. This amounts to the confirmation of a prediction. Under what condition would $Q(x)$ be nonprojectable, by contrast? This condition cannot be that $Q(x)$ is not exhibited in *B* at all: After all, we can be sure that we can express *B* as the sum of $Q(x)$ and some noise level or other. Rather, $Q(x)$ is nonprojectable presumably if the noise level with which it is exhibited in B is higher than n%. This amounts to the disconfirmation of a prediction.

These outcomes may represent interesting discoveries about datasets A and B. They do not, however, indicate that pattern $P(x)$ has any greater claim to correspond to a structure in the world than $Q(x)$. The fact that $P(x)$ is exhibited in both A and B with the same noise level, while $Q(x)$ is exhibited in B with a higher noise level than in A, has no bearing on the ontological question of which of them are evidence of structures in the world and which are not. It merely causes us to revise our estimate of the noise level with which $Q(x)$ is exhibited in B.

The concept of projectability is useful for keeping track of the scope of our claims: We often want to know whether a claim that we have established about a dataset A—namely, that a pattern $P(x)$ is exhibited in it with a noise level of n%—holds also for a subsequent dataset B. As a criterion for distinguishing real from other patterns, however, it would serve only if we had reason to believe that patterns were evidence of structures in the world if and only if the noise level with which they were exhibited in B were no higher than that with which they were exhibited in A. This suggestion is implausible: It amounts to tying reality of patterns to an arbitrary noise threshold—in this example, n%.

The projectability criterion, in short, offers no solution to the problem of distinguishing real patterns from other patterns. As Dennett noted, patterns exhibited in empirical datasets with different noise levels can all support successful predictions. Absent some reason for thinking that a certain noise level picks out patterns that have more claim to correspond to structures in the world, all patterns have equal such claim.

The irrelevance of the temporal sequence of making and verifying predictions to the ontological question is also evident in the following example. Consider an ultimate dataset, U, encompassing all observations made in science. Presumably, U provides the most reliable empirical base for establishing which structures the world contains. However, U exhibits all possible patterns, each with a certain noise level. The real patterns hypothesis assumes that some of the patterns displayed in U correspond to structures in the world, while others do not. But what distinguishes patterns belonging to these two categories? Clearly, the historical sequence in which we ascertained the noise levels associated with patterns in U is irrelevant for resolving this question.

Next, consider the criterion of economy of description, which associates real patternhood with a pattern's ability to encode data economically. Both Dennett and structural realists have suggested this as key to distinguishing real patterns from bogus ones. Yet, this criterion too falls short, or so I argue.

An economical description of a dataset typically consists of two items: A concise formula that generates a regular pattern and a bit-by-bit record of the discrepancies between the data and this pattern. Expressing the data in terms of these two items is typically more economical than a pure bit-by-bit transcription if the pattern accounts for a large part of the variance of the data, so the bit-by-bit element need contain less information.

The claim that a pattern's ability to compress a dataset reveals whether the pattern is physically significant is implausible, for the following reason. The effectiveness of a particular pattern in compressing a given dataset depends on two things: The form of the pattern and the bit-by-bit value of the dataset. The latter depends partly on the amount of noise that it contains, however. The amount of noise in a dataset is unrelated to the properties of the pattern. This suggests that the degree to which a pattern compresses a dataset tells us very little about the properties of the pattern.

Suppose we collect a certain empirical dataset, $E(x)$, in an experiment. This exhibits two patterns: $P_A(x)$, a real pattern that is caused by a structure in the world, A, and $Q(x)$, a bogus pattern that is not caused by anything. Suppose the most economical compression of $E(x)$ is in terms of $P_A(x)$—that is, as the sum of $P_A(x)$ and its associated noise term. In this case, economy of description would correctly designate $P_A(x)$ as a real pattern, and $Q(x)$ as nonreal.

But now consider the following. The noise term with which $P_A(x)$ is exhibited in the dataset is, of course, $E(x) - P_A(x)$. This noise term is the effect of various causal structures in the world distinct from A. Call this combination of structures B. We can rearrange the causal chains in the world in such a way that structure A interacts not with B but with a new set of causal factors, B', to generate a second empirical dataset, $E'(x)$, in a new experiment. Structure A in the world still causes its distinctive pattern, $P_A(x)$, which is displayed in the new dataset, $E'(x)$, with noise term $E'(x) - P_A(x)$.

By carefully choosing the new set of causal factors, B', we can determine the magnitude and distribution of the new noise term, $E'(x) - P_A(x)$, as we please. This means that we can ensure that $P_A(x)$ accounts for less of the variance of $E'(x)$ than $Q(x)$ does. As a consequence, we can ensure that the most economical compression of the new empirical dataset, $E'(x)$, is in terms of $Q(x)$, and no longer of $P_A(x)$. $Q(x)$ was, as we stipulated, a bogus pattern. This shows that economy of description is not a reliable criterion of evidential significance: It designates a pattern as having evidential significance, or as corresponding to a structure in the world, when exhibited in one dataset with one noise term, but as lacking evidential significance when exhibited in another.

In essence, which pattern economy of description picks out as real depends on the overall value of a dataset, and thus on the amount of noise that it contains. That is not a desirable characteristic for a criterion of real patternhood.

In passing, it is worth noting that this argument establishes more generally that a pattern's evidential significance cannot depend on the properties of the noise term with which it is exhibited in empirical data. No noise term, irrespective of its properties—in particular, its magnitude and whether it contains patterns of its own—can be grounds for ruling a pattern as lacking evidential significance. This is understandable: The properties of something that is only contingently associated with a pattern—the noise with which it is exhibited in a given dataset—cannot determine whether it corresponds to a structure in the world.

In conclusion, projectability and economy of description do not provide adequate criteria for distinguishing patterns that correspond to structures in the world from those that do not. The properties suggested as features distinguishing real patterns are, in fact, shared by all patterns found in empirical data (Beni 2017, 291–297).

Of course, projectability and economy of description are not the only conceivable ways for distinguishing real from bogus patterns. Other options include consilience, the idea that patterns in different datasets that constitute evidence of the same structures in the world are real, and grounding, the notion that we are justified in regarding a pattern in a given dataset as real in the light of discoveries about other patterns in previous datasets. I refer to other work for critical scrutiny of these further proposals (McAllister 2010, 808–809; 2011, 79–81).

7.7 All Patterns Are Real

Half-way through this chapter, we came to a pivotal choice. The fact that, mathematically, any empirical dataset could be expressed as the sum of any possible pattern of the appropriate length and an associated noise term gave us two options. The first was to say that only some patterns in data were evidence of structures in the world—what I called the bipartition thesis; the second was to say that all patterns were. We have thoroughly examined the first option. I argued that a distinction between patterns that did and that did not correspond to structures in the world would be otiose without a criterion for recognizing the two sets of patterns in empirical data. I contended that no such criterion existed, especially not in the forms envisaged by proponents of the real patterns hypothesis.

Now, let us consider the alternative view: That all patterns in data constitute evidence of structures in the world.

There are compelling grounds for supporting this conclusion. A pattern is simply an additive component of an empirical dataset. Since the world is the cause of the entirety of the variability of empirical data, it follows that any additive component of a dataset must be caused by some subset of causal factors in the world. In other words, regardless of how we pick out a pattern, there must be something in the world that generates it; if nothing did, then the magnitude of the dataset as a whole would also be undetermined. While the specific subset of causal factors in the world responsible for a given pattern may be unknown or unclear, it is undeniable that such factors exist, and that the given pattern is a trace of them in the data. Thus, any pattern exhibited in empirical data constitutes evidence of some structure in the world.

To illustrate, recall our dataset on sea surface heights. The totality of causal factors acting on the oceans ensures the overall measurement outcomes of sea surface heights that oceanographers obtain. Now take an arbitrary pattern, *S*, in those data: There exists a subset of causal factors affecting sea water that determines the magnitude of the additive component of the data that corresponds to *S*. Pattern *S* is, therefore, evidence of that subset. If that subset did not exist, the overall measurement outcomes would be undefined.

Some might object to this view by noting that science has yet to pinpoint causal factors behind the vast majority of patterns that we might pick out in empirical data. That is true, but unsurprising given the current, incomplete stage of scientific knowledge and the selective cognitive and practical interests of investigators. Another objection might be that there is no evidence of a physical factor responsible for an arbitrary pattern that is discernible in empirical data. This objection is misguided, however: The pattern itself is that evidence. Empirical evidence of structures in the world never consists of anything more than patterns in data. For instance, the patterns identified by oceanographers in sea surface height data are themselves the only empirical evidence of the physical processes responsible for them.

We may further test this view by scrutinizing one of its implications. If every component of an empirical dataset constitutes evidence of structures in the world, then even what is designated as the noise term for a particular research project must contain information about some structure in the world, albeit not one pertinent to that project's focus. In other words, noise in one context must be signal in another.

I believe this to be the case. What we call noise is also the result of causal influences in the world. We may label these as extraneous causes, perturbations, accidents, or confounding factors, and characterize them as random, irregular, incidental, or ephemeral. For all that, these factors are genuine components of the world's causal structure, and analyzing noise can unlock information about them. Indeed, scientists often make advances by reanalyzing components of empirical data previously dismissed as noise and extracting valuable information about further structures in the world. This suggests that the partition of a dataset into pattern and noise is contingent on the aims of a particular research project and corresponds to no objective difference among causal factors in the world, confirming that all discernible patterns in empirical data correspond to structures in the world.

There are numerous historical examples of the discovery of useful information in what was previously classed as noise. To give just one instance, astrophysicists using ground-based observation to study star spectra disregarded the absorption lines caused by atmospheric ozone in spectrographs as noise in their data: These lines were not relevant to the study of stellar spectra. However, R. Elizabeth Griffin (2005) reanalyzed those data to answer a question in a different branch of science: How ozone levels in the atmosphere varied during the twentieth century. What was noise for astrophysicists studying stellar spectra became evidence for climatologists studying atmospheric pollution.

While all patterns that datasets exhibit correspond to structures in the world, scientists are under no obligation to investigate them all. Most scientists will choose, from the infinitely many patterns available in a given empirical dataset, just one or two that they designate as the evidence in their research project, and attribute the rest of the variability to noise. They may choose these patterns on any criteria they please, of course. In most cases, they will pursue their cognitive and practical interests by choosing patterns on two criteria: A certain mathematical form, such as a polynomial function corresponding to a straight line or a harmonic function corresponding to a sinusoidal curve in a graphical plot, and a certain noise level, which picks out patterns that follow the variance of the data more or less closely. Patterns exhibited with lower noise often correspond to smaller-scale structures, while those exhibited with higher noise correspond to structures on larger scales. Variants of these criteria are also embedded in our sensory and cognitive apparatus, as well as in many observational instruments.

Of course, the pragmatic criteria on which scientists choose which patterns to study do not determine whether a particular pattern has or lacks evidential significance. Pragmatic criteria do not ground metaphysical claims, as we saw above. Indeed, investigators would not need to base choices of patterns on preferences for a certain mathematical form or noise level if it were clear that just some patterns in a given empirical dataset had evidential significance.

I leave open the further question how many distinct structures the world contains. The claim that all discernible patterns in empirical data are evidence of structures in the world does not, by itself, determine how many different structures these are. One possibility is that a single structure in the world gives rise to multiple patterns in data. If so, then the argument presented here would still be compatible with the claim that the world contained a finite and perhaps even relatively small number of distinct structures.

I doubt this view: I prefer to adhere to the principle that our sole way of individuating structures in the world is through patterns in empirical data. Distinct patterns are evidence of distinct structures. This leads me to hypothesize that there are as many distinct structures in the world as there are patterns in empirical data—a metaphysical perspective that I have termed "radically polymorphous" (McAllister 2011, 84–86). This further hunch, however, extends beyond the scope of the present chapter.

7.8 Conclusions

The real patterns hypothesis is a bold attempt to link the evidential content of empirical data, and by extension the ontology of science, to properties of patterns within those data. The hypothesis proposes a distinction between real and nonreal patterns to explain how scientists mine empirical data for evidence about structures in the world.

However, as I have argued, the hypothesis meets a critical problem: There is no criterion to distinguish real patterns from nonreal ones. Once we acknowledge that evidence of structures in the world manifests itself as patterns in empirical data, the very nature of patterns resists any attempt to limit evidential significance to just a subset. This opens up an unexpected ontological vista: All patterns have equal claim to reflect structures in the world. Every discernible pattern in empirical data holds insights into the structure of reality. In the terminology of the hypothesis, all patterns are real.

Some might see this conclusion as a refutation of the real patterns hypothesis. I, however, prefer to see it as complementing the hypothesis and bringing it to its logical fulfillment: The intuition that evidence about structures in the world comes in the form of patterns in data is now extended and placed on a stable footing. This expanded view invites further exploration of what the full spectrum of patterns entails for our understanding of the universe.

7.9 Acknowledgments

I discussed these ideas at the workshops: "Real Patterns," Universitat de Barcelona, December 2018, "Real Patterns in Science and Cognition" (online), Santa Fe Institute, March 2022, and the 47th Annual Philosophy of Science Conference, Inter-University Centre Dubrovnik, April 2022. I am grateful to the organizers and participants for these opportunities. My thanks go also to the volume editors and three unnamed referees for critical remarks, which prompted me to reconsider several key issues.

References

Beni, M. D. 2017. "Structural Realism, Metaphysical Unification, and the Ontology and Epistemology of Patterns." *International Studies in the Philosophy of Science* 31 (3): 285–300.

Bogen, J., and J. Woodward. 1988. "Saving the Phenomena." *Philosophical Review* 97 (3): 303–352.

Dennett, D. C. 1991. "Real Patterns." *Journal of Philosophy* 88 (1): 27–51.

Griffin, R. E. 2005. "The Detection and Measurement of Telluric Ozone from Stellar Spectra." *Publications of the Astronomical Society of the Pacific* 117 (843): 885–894.

Hanson, N. R. 1958. *Patterns of Discovery: An Inquiry into the Conceptual Foundations of Science.* Cambridge University Press.

Hon, G., and B. R. Goldstein. 2005. "How Einstein Made Asymmetry Disappear: Symmetry and Relativity in 1905." *Archive for History of Exact Sciences* 59:437–544.

Ladyman, J. 2017. "An Apology for Naturalized Metaphysics." In *Metaphysics and the Philosophy of Science,* edited by M. Slater and Z. Yudell. Oxford University Press.

Ladyman, J., and D. Ross. 2007. *Every Thing Must Go.* Oxford University Press.

McAllister, J. W. 1997. "Phenomena and Patterns in Data Sets." *Erkenntnis* 47 (2): 217–228.

McAllister, J. W. 2010. "The Ontology of Patterns in Empirical Data." *Philosophy of Science* 77 (5): 804–814.

McAllister, J. W. 2011. "What Do Patterns in Empirical Data Tell Us About the Structure of the World?" *Synthese* 182 (1): 73–87.

Millhouse, T. 2021. "Compressibility and the Reality of Patterns." *Philosophy of Science* 88 (1): 22–43.

Motter, A., and D. Campbell. 2013. "Chaos at Fifty." *Physics Today* 66 (5): 27–33.

Petersen, S. 2023. "Multiple Patterns, Multiple Explanations." In *Conjunctive Explanations: The Nature, Epistemology, and Psychology of Explanatory Multiplicity,* edited by J. N. Schupbach and D. H. Glass, 38–48. Routledge.

Pugh, D., and P. Woodworth. 2014. *Sea-Level Science: Understanding Tides, Surges, Tsunamis and Mean Sea-Level Changes.* Cambridge University Press.

Suñé, A., and M. Martínez. 2021. "Real Patterns and Indispensability." *Synthese* 198 (5): 4315–4330.

Wong, C. Y. S., and J. A. Gamon. 2015. "Three Causes of Variation in the Photochemical Reflectance Index (PRI) in Evergreen Conifers." *New Phytologist* 206 (1): 187–195.

Worrall, J. 1989. "Structural Realism: The Best of Both Worlds?" *Dialectica* 43 (1-2): 99–124.

8 Real Patterns in Physics and Beyond

David Wallace

8.1 Introduction

Dennett's seminal "Real Patterns" (Dennett 1991) is first and foremost an intervention in the philosophy of mind: Its goal is to defend Dennett's (1981; 1989) intentional stance theory of content from the supposed dilemma that either beliefs are real (in which case there had better be some concrete *things* in the world, some collections of neural tissue or similar, with which they can be identified) or they are not (in which case they don't exist and so cannot play a role in any science of the mind worth the name). In "Real Patterns," beliefs are real patterns in the behavioral dispositions of believing systems, discernible through the intentional stance: abstracta to be sure, but useful abstracta, abstracta that earn their keep through the predictive and explanatory work they do and the shortcuts they allow us to take compared to the impossibly demanding and in any case uninformative task of working out what those systems would do *qua* systems of concrete matter. Abstracta that useful ought to be considered along with paradigm physical entities like electrons as part of what science tells there is in the world.

But science—even physics—is not a monolith, and once we dig deeper into physics, these "paradigm" physical entities start looking more problematic, and "concrete" matter starts looking less concrete. Physics describes the world on many levels, from quantum field theory, through the quantum-mechanical descriptions of nuclei and electrons, up to the quasi-classical world of molecular physics and thence to the comparatively familiar world of fluids and solids. And at those different levels, its ontology looks very different—at one point fields, at another particles, at still another continuous matter. How those various ontologies are related—how ontology at one level might be emergent from different ontology at a lower level—is not at all obvious or simple, and over the last twenty years I and others[1] have deployed the real-patterns idea inside

1. See, e.g., Wallace (2003; 2010; 2012); Seifert (2023); Franklin (2024); Franklin and Robertson (2024); Ladyman and Ross (2007); Ross (2000). (Not all these authors agree: Ladyman and Ross,

physics itself, to elucidate how an ontology of one kind can be emergent from an ontology of a quite different kind.

This chapter is about that project, and its relation to another major theme in recent philosophy of physics, *ontic structural realism*, the proposal (at least as I read it) that the Quinean project of reading ontological content off our best scientific theories assumes that those theories can be faithfully understood as sets of claims in natural language, and that once this claim is abandoned, the very idea of looking for the ontology of a theory becomes at best a second-order activity. In section 8.2, I spell out the ontological puzzle physics poses, and in section 8.3, I explain how the real-patterns idea transforms but does not fully resolve the puzzle. In section 8.4, I describe (my version of) structural realism, and then in section 8.5, I relate it back to real patterns and to the project of ontology. In section 8.6, I briefly consider how these ideas might find application beyond physics.

The physics I discuss in this chapter is well established, and I do not attempt to give original references.

8.2 Emergence of Ontology in Physics: Molecules and Fluids

My concern in this chapter is with intertheoretic relations *within* physics, and that might seem puzzling, even perverse. Don't the more interesting questions concern the reduction of other sciences *to* physics? And insofar as there are reductions inside physics—of atoms to electrons and nuclei, of thermodynamics to statistical mechanics—aren't they supposed to be the paradigm cases of well-understood reduction, to contrast with the knottier cases in the special sciences? But if what makes a science "special" is that it lacks universality of scope, almost all of physics is a special science—physicists study stars, or metals, or plasmas, or lasers—and much of the work of physics is constructing one intertheoretic relation or another.

Those relations are far harder to understand philosophically than is sometimes recognized in philosophy: The reduction of thermodynamics to statistical mechanics, in particular, is hugely controversial and comprises one of the central problems of modern philosophy of physics.[2] But for all that they are the simplest intertheoretic relations we find in the sciences, and for that reason they serve as test beds for any general account of intertheoretic relations. If we cannot understand how theories are related even within physics, we have

in particular, interpret real patterns not as a between-levels relation from lower-level to higher-level ontology, but as a within-levels relations between ontology and phenomena.)

2. See, e.g., Sklar (1993), Albert (2000), Frigg (2007), Myrvold (2022), and Wallace (2023).

little hope of understanding the more complex cases; if we can understand the physics case, perhaps we can learn wider lessons.

Let's consider a comparatively simple example.[3] Our system will just be a sealed box, a few centimeters across, containing some fluid—chlorine gas, water, molten iron, the details will not matter. We can distinguish (at least) two levels of description of that system:

The interatomic level: The system is a collection of order 10^{24} atoms (or molecules), characterized by the masses of those atoms and by the functional form of the fairly-short-range forces by which neighboring atoms interact with one another. In some cases we might need to engage with the substructure of the atoms, but in many cases the interactions *between* atoms occur at a sufficiently different energy scale from the interactions *within* atoms that we can get away with treating them as elementary. A full description of the system requires the 6×10^{24} real numbers that represent the positions and velocities of each atom; the evolution of the system is governed by Newton's laws, supplemented by the details of the particular interatomic interactions that apply.[4]

The fluid level: The system is a continuous fluid, characterized by parameters like specific heat capacity and viscocity and by how its density, pressure, and temperature vary across the box. A full description of the system requires several smooth functions, encoding those variation; the evolution of the system is governed by the Navier-Stokes equation of fluid dynamics.

For the moment, let's pretend that the interatomic-level description is the most fundamental description we have, and disregard the realms of atomic and nuclear and particle physics.

This is about as simple a case of an intertheoretic relation as we can imagine. If we can't understand how *fluids* emerge from atoms, our prospects for understanding how *beliefs* emerge from them look dismal. And yet already there are subtleties. Consider: What's in the box? "Atoms," is the obvious answer: Our best theory of the box quantifies over atoms, asserts that atoms exists, so (following Quine [1948]) if we accept that theory, we accept that there are atoms.

3. For technical discussion of the relevant physics, see, e.g., Balescu (1997) and references therein.

4. I assume here that the system can be understood classically, without the use of quantum mechanics except perhaps to calculate the form of the interatomic interaction. This is not always valid technically and is conceptually subtle even when it is technically permissible, for reasons connected to the quantum measurement problem (see Wallace [2016] for discussion), but including those subtleties would if anything strengthen the points I wish to make here, so I disregard them for simplicity.

But isn't it also true that there is a fluid in the box? In the interatomic-level description there are no fluids to be found: only atoms and the void. Fluids are not quantified over according to the interatomic theory, and that (we're pretending) is our *fundamental* theory of the box, so (again following Quine) there *is* no fluid in the box. That seems absurd: The empirical evidence for water is pretty conclusive. Indeed, the empirical evidence for the *theory* of fluids is pretty solid, and that theory does quantify over fluids. So how do we reconcile this with the absence of fluids at the interatomic level? The metaphysics-of-science literature has offered three main answers.

The first is *disunity*: There is the fluid theory, valid for fluids, and the interatomic theory, valid for the interactions of molecules; both are well evidenced by the data; both apply in their particular domains; perhaps there are overlaps in their applicability, but there is no hierarchy of fundamentality, no reason but prejudice to regard the "lower-level" theory as any more of a guide to ontology than the higher-level theory. Theses of this kind (of varying strengths) have been quite popular in the philosophy of the special sciences, for instance in the work of Cartwright (1999) and Dupré (1983) (often drawing inspiration from Fodor [1974] though Fodor's own thesis is less ontologically radical): They are tied to the denial of reductionism, of the idea that the generalities and models of the special sciences could in any feasible way be derived from those of physics.

For myself, I find those claims overstated, and based on an overly reductive understanding of reductionism itself (cf. Dennett 1995, 80–83). Among other observations, the special sciences are replete with physics ideas: Biochemists discuss molecular bonds (cf. Hermida and Ladyman 2022); biophysicists study the physics of animal movement; organic chemists rely on thermodynamics; doctors use CT scans and positron emission tomography. But what makes them even plausible is that reduction between physics and the special sciences is at best incomplete and in general aspirational. We may or may not have good reasons to think that cognition, say, reduces to physics in principle, or even that we will someday obtain that reduction ourselves, but today we have, at most, suggestive fragments of that reduction.

The project of reducing fluid dynamics to interatomic physics, however, is far from aspirational: Huge parts of it have actually been completed. We know how to derive the Navier-Stokes equations from the microdynamical equations; we know in principle how to calculate quantities like specific heat capacity and viscosity from the microdynamics; in many cases (dilute gases most completely, but not only there) we have actually done those calculations and got answers that match the data. It's not *easy*, and the devil is sometimes in the details,[5] but there is not a lot of room to deny the claim that the fluid-level physics is derivable, in the appropriate regime, from the atomic description. And then it seems inexplicable how, if we actually have a well-controlled account of the mathematical relation between the two theories, it could still be the case that their ontologies are just disconnected.

The second answer, *composition*, embraces reduction in the fullest sense: The fluid actually is present in the lower-level description, because the fluid just *is* all of the atoms together. At the formal level, this means supplementing our lower-level theory with a *mereology*, a logic of whole and part, so that as well as quantifying over the individual atoms we quantify over the various mereological sums of atoms. The fluid is then identified with one of those mereological sums—the sum of *all* the atoms—and the spatial parts of the fluid with the mereological sums of subsets of the atoms.

Composition meshes naturally with the traditional (Nagelian) approach to reduction. There (Nagel 1961), intertheoretic reduction is logical deduction of one theory from another with the aid of "bridge laws," auxiliary axioms that coordinate the terms of one theory with those of another—for instance, "The mass of the fluid in region R is Nm iff there are N atoms each of mass m in region R." For that notion of reduction genuinely to *reduce* one theory to another, those bridge laws have to be more than mere contingencies: They must be something like identities. (Lewis [1970] is perhaps the best-known working out of this idea.) And then something like mereological composition seems unavoidable: There need to be things in the ontology of the microscopic theory that we can treat as identical to the fluid and its parts, and an austere ontology of particles alone contains no candidates. The overall package is undeniably elegant, and meshes well both with the formal methods of analytic philosophy and with our intuitive understanding of composition.

Its mesh with actual physics is less impressive. We might expect to find mereology—or some rough-and-ready version of it—in the actual mathematics by which fluid dynamics was derived, but it is not at all apparent there. (In the clearest derivations I know, the fluid gets identified with some subcollection of the lowest-frequency Fourier modes of the particle distribution.) And the mereological sum of the molecules seems quite different in nature from the

5. See Batterman (2021) for philosophical discussion of some of the subtleties, and of the need in many cases to establish a robust "mesoscale" understanding of a continuum system. It is also generally the case that the higher-level theories are irreversible in the statistical-mechanics sense, whereas the lower-level theories are reversible, so that somewhere a time-asymmetric assumption has to be introduced; a full understanding of the reduction therefore also requires some engagement with the vexed question of the arrow of time; see, e.g., Albert (2000), Price (1996), and Wallace (2013).

fluid posited by fluid dynamics—the former consists of isolated points scattered through empty space; the latter is a smooth continuum. The situation only gets worse in more esoteric regimes of physics. The statement that a proton is composed of three quarks, for instance, is somewhere between a forgivable heuristic in popular science and a story we tell children; quarks themselves are identified with certain excitations of an underlying field, *very roughly* in the way that an ordinary wave is identified as a moving disturbance of a fluid. There is no obvious logic of composition identifiable here that transcends the messy details of the specific cases; indeed, those specific cases often mix up ontological categories, "identifying" a higher-level object with something like a property of the lower-level objects.

If neither disunity nor composition offer a scientifically appropriate account of higher-order ontology, that brings us to the third answer: *instrumentalism.* Perhaps there *is* no higher-order ontology; it is just convenient to pretend that there is. The instrumentalist need not worry about establishing some precise connection between higher-level and lower-level ontology, because really there is only lower-level ontology—but it is sometimes pragmatically helpful to pretend, to act as if there is high-level ontology. That pretense can make it simpler to carry out various scientific tasks, to predict and explain the phenomena, but ultimately those phenomena are only phenomena at the high level. The instrumentalist about fluids will say that "fluid" is just a convenient way of talking, but that really there are just atoms in certain configurations—"arranged fluid-wise," to borrow a term from the metaphysics literature.

Applied to intentional states, this is already radical. Applied to physics, it becomes dizzying. Once we drop our pretense, we remember that the theory of classical atoms and molecules is not fundamental either. It rests on the quantum mechanics of the subatomic, which rests on the layered quantum field theories of particle physics. Even the deepest of those is not fundamental: The Standard Model of particle physics is an effective theory, a low-energy, large-scale approximation to something about which we can only speculate (string theory being the furthest developed of those speculations). To be an instrumentalist about "higher-level" physics is to be an instrumentalist about every scientific entity ever described in an empirically confirmed theory. Indeed, it is to be an instrumentalist about scientists themselves, about their—our—experiments and language. Even space and time are generally expected to be emergent consequences of a deeper, non-spatiotemporal physics—to the instrumentalist, they too don't really exist.

At this point, instrumentalism about the high level starts to look incomprehensible, if not incoherent, and we seem forced to admit at least *some* higher-level ontology. But how much? Suppose two theories agree that atoms exist, but disagree about whether fluids do; what, actually, is the form of the disagreement? It cannot be stated using the vocabulary of the interatomic theory: That theory, in itself, requires no mereology. It cannot make a difference to anything stated in the language of interatomic ontology, for the same reason. What *does* it make a difference to? The question cannot be answered naturalistically: Science is silent on whether fluids really exist or are just an indispensable-in-practice fiction.[6] We seem to have been led from a quintessentially naturalistic problem—given that systems are described on various levels by various different theories, how do we think about the relation between those theories' respective ontologies?—step by step into the deepest jungles of analytic metaphysics.

8.3 Real Patterns and Dennett's Criterion

The last section is a sort of Rorschach test for philosophers. For some, they illustrate exactly why metaphysics is indispensable, for all that neo-positivists and other philistines might claim otherwise: Think clearly enough about any deep problem in philosophy of science and eventually you will have to get your metaphysics straight, and in doing so will have to engage in the practice of metaphysics. For others—like myself and, I imagine, Dennett—the fact that we seemed to be led so deeply away from anything naturalistic is an indication that something went badly wrong in our setup of the problem, and we should work out what it is.[7]

At this point "Real Patterns" comes as a breath of fresh air. In Dennett's proposal, higher-level entities are not literally *composed* of lower-level entities, but nor are they simply instrumentalist fictions: Rather, they are patterns—or, in terms more natural to a philosopher of physics, *structures*—in the distributions and dynamics of the lower-level entities. The fluid is not a mereological sum of the atoms in the box: It is a pattern in their behavior, a pattern elucidated by the mathematical derivation of fluid dynamics from interatomic physics, and it is a *real* pattern, and not just a "mere" pattern, because it is such a powerful, explanatory, predictive pattern. There is a flexibility, a situation-dependence, to the notion of pattern not shared by something as rigid as composition: A fluid can be *truthfully* described as continuous simply because a continuum is the best way to describe the pattern that fluid dynamics discerns

6. If you doubt this, try asking a few fluid dynamicists whether fluids really exist or are just a useful way of talking about atoms. Then try the analogous question about atoms themselves.

7. To avoid being misunderstood: I don't want to suggest that naturalistically inclined philosophers cannot benefit from engagement with modern metaphysics. My own development of these ideas has been greatly helped by (admittedly critical) engagement with it: I'll note in particular the influence of Sider (2020; 2024).

in the microphysics. Similarly, *pattern* need not respect ontological categories: The pattern (an object) might be realized not in the objects of the lower-level theory but in its properties and quantities.

Once you understand the idea, and once you learn a little physics, you begin to see real patterns everywhere. Light is a traveling pattern in the excitations of the electromagnetic field (just as, in "Real Patterns," a glider is a traveling pattern in the excitations of the cells in the Game of Life). Physicists describe the vibrations in a crystal by treating it as a gas of phonons, the quanta of sound: These phonons are patterns in the movements of the atoms in the crystal. Electrons are patterns in the excitations of the quantum fields of the Standard Model.

In previous work (Wallace 2003; 2012) I called the general principle here

Dennett's Criterion: A macro-object is a pattern, and the existence of a pattern as a real thing depends on the usefulness—in particular, the explanatory power and predictive reliability—of theories which admit that pattern in their ontology.

The theory of fluids, for example, is de facto indispensable in studying large-scale features of certain molecular systems—various of their features would be in principle inexplicable and in practice impossible to calculate without using fluid dynamics—and so fluids are real.[8]

A real-patterns approach does *drastically* better at capturing the way physicists themselves talk about ontology than do theories based on composition or disunity (while avoiding the nihilism of a thoroughgoing instrumentalism toward physics). In physics as in philosophy of mind, real patterns offer the promise of a middle way between instrumentalism and realism, which closely fits the practice of physics. Physicists are extremely permissive about introducing new higher-level ontology, and then talking about it as robustly as they did their old ontology: They are not concerned about any particular way that the new ontology is related to the old but they do require it to be part of a theory derived or constructed from the lower-level physics, and they do require it to be *useful*.

(In my own original use of Dennett's criterion, I used physics practice to justify the claim that ontology *in general* works this way in physics, and then applied it to one specific controversial case: The "branches" or "worlds" that

8. There is a close analogy here to Dennett's discussion of the indispensability of the intentional stance in Dennett (1989, 225–227): In practice it is infeasible for us to predict what intelligent agents will do except through the intentional stance, but even if alien super-scientists did have the ability to make atom-by-atom predictions of what a human being would do, they would still be missing something explanatory if they didn't know that the intentional stance licensed much simpler predictions.

appear in the Everett (Many-Worlds) interpretation of quantum mechanics. The idea was that the "worlds" are real patterns in the evolving quantum system, so that according to the general ontological criteria that physicists tacitly use, the worlds really exist, as higher-order, non-fundamental ontology.)

That said, while I think something along the lines of the real-patterns approach is correct, by itself it still leaves some metaphysical questions unanswered. For one, while the idea of a pattern is certainly intuitive, it is harder to pin down what it means to quantify over patterns in an underlying ontology, especially when that underlying ontology is as alien and unintuitive as some of those found in physics. For another, Dennett distinguishes between *real* patterns and *mere* patterns, as (in effect) does my phrasing of Dennett's criterion. But then what does it actually add to say that a pattern is *real*? It doesn't mean that it *exists*, in the Quinean sense: All the patterns exist, in the sense that we're quantifying over them when we identify some as mere and some as real. Nor will it do to get pragmatic and say that it is useful to treat some patterns *as* real: Nothing in scientific practice will distinguish the scientist who does this from the scientist who simply *uses* the "real" patterns without applying the label "real" to them.

Perhaps we should instead read Dennett's criterion as some sort of abstraction principle: "Pattern X exists iff the underlying objects are arranged X-wise and positing X is . . ." (In my own version of Dennett's criterion, the . . . is filled in by some statement of explanatory and predictive power). If so, what is the status of that principle? It is tempting to read it pragmatically, as a statement about when it is useful to use language that way—but that gets us to quite an instrumentalist conception of higher-order ontology, one that becomes even more uncomfortable when it is not just beliefs and desires but solid matter itself that is to be understood as a real pattern. If we try to read it realistically, we are in the uncomfortable position of having a criterion for *real ontology* which itself turns to some degree on pragmatic considerations. If there are alien physicists[9] with quite different pragmatic goals from ours, would quite different patterns not be real to them? Does that kind of interest-relative ontology even make sense? And where does it leave scientific realism?

Worries along these lines lead real-pattern advocates like Ladyman and Ross (2007) to fill in the . . . with something a little more objective: informational compressibility from a physically possible perspective. But then what is the status of that principle—is it some metaphysical a priori truth, and if so, how do we come to know it? If two fans of the real-patterns idea fill in the blanks a little differently, how is that disagreement to be resolved? Most of all, what really

9. Thanks to an anonymous reviewer for suggesting this nice example.

distinguishes the person who holds the abstraction principle to be *true* from the person who regards it as a mere fiction? The impasse between compositional and instrumentalist approaches to higher-order ontology threatens to recur in more naturalistic guise.

There is something frustrating about this whole dialectic. Again, our starting point is as naturalistic a philosophical question as one could wish—what is the world like, given that we describe it with different physical theories at different scales—but our attempts to answer it keep ensnaring us in metaphysical tangles that naturalistic methods cannot straighten out. There must still be something wrong with how we are thinking about the problem; the real pattern idea is a major step forward, but it can't be the whole of the solution.

8.4 Mathematics and the World

Let's step back a moment. We are trying to understand the relation between what some system, some part of the world, is like according to a higher-level theory (in our example, fluid dynamics) and what it is like according to a lower-level theory (like interatomic physics) that purports to describe the same system on different scales. But at least in the context of physics, it is odd that there can even be a substantive question—after all, *ex hypothesi* we understand how interatomic physics describes the system, *and* we understand how fluid dynamics does, *and* we understand the mathematical relation between the two systems. How can there be anything substantive left to understand?

To answer this, we need to look more closely at what these theories of physics actually are. So far I have been writing as if a scientific theory was something like a description of the world (or a part of it) in language: That is, a theory *says* what things there are in the world, and what properties they have and relations they stand in, and how all this changes over time. (It's clear that this is the notion of theory that Quine, at least, has in mind: something given in the first instance in a natural language like English, but which at least in idealization we can imagine being reformulated in the predicate calculus.) But at least in physics, scientific theories aren't like that any more, if indeed they ever were. Physics theories are highly mathematized, and the relation they bear to the systems they describe is not best thought of through linguistic notions of truth and reference and satisfaction but by some more general notion of *representation*.

Why does this matter? Because the description of the system given in the mathematics is typically more coarse-grained than a purely linguistic description would be. Consider again the fluid: It is characterized in part by a function that represents the density at each part in the fluid, but there are two ways to think about density:

Density fundamentalism: Density is just a primitive feature of pointlike parts of a fluid; integrating it over regions of the fluid gives a derivative quantity, mass.

Mass fundamentalism: The real primitive is mass: Every[10] nonzero-volume part of the fluid has a mass; the mass of a disjoint sum of parts is the sum of their separate masses; it is density that is the derivative quantity, defined as the limit of the ratio of mass to volume as the latter approaches zero.

Try to formalize the theory of the fluid in predicate logic and you will have to make a call on questions like this: The two choices lead to different logical vocabulary and, indirectly, to different metaphysical commitments. But mathematically they are interchangeable: Fluid dynamics doesn't care, and neither do fluid dynamicists. Or more precisely: They might care about which way of describing density is clearer and more useful, but not about which is true.

Examples like this multiply and become more radical as we move to parts of physics further from the manifest image. Consider a classical field theory like electromagnetism. Mathematically (and in the absence of charged matter) the theory represents the electromagnetic field by a pair of vector field, assigning an electric and a magnetic field vector to every point in space. Should we interpret this as saying that there is an extended entity—a field—whose parts occupy regions of space and have field strengths as their fundamental properties? Or is "field" just a façon de parler: Are "field strengths" new properties of the points of space? The two interpretations differ radically in ontology: According to one, there are continuum many new entities, the various parts of the field; according to the other, there are no non-spatial entities at all—and yet the mathematics of electromagnetism makes no distinction between them.

Following Wallace 2022, a *predicate precisification*[11] of a mathematically stated theory of physics is an account *in natural language* of what a system is like according to that theory. Precisifications need to commit on various questions of ontology and ideology that the theory itself is silent on, but "precise" here is not a virtue term: A predicate precisification commits to more about the system than physics itself licenses. That is: A theory of physics underdetermines its own description in natural language.[12]

10. Here I elide some measure-theoretic technicalities: It is not possible to define a volume measure on *every* subset of the continuum. See, e.g., Halmos 1950.

11. I concede that this is not a beautiful phrase. I welcome suggestions for improvement, but there are false friends lurking: "Predicate description" sounds more elegant but unavoidably suggests that what we describe in natural language is the theory itself (a mathematical structure) rather than the system that the theory represents.

12. There is an interesting analogy with Dennett's (1989) distinction between *beliefs* (intentional states characterized through the intentional stance, applicable to any system that can usefully be ascribed them) and *opinions* (commitments made in language, available only to language users).

It is tempting to think that some predicate precisification of a successful theory must be true, so that scientific theories represent the world by virtue of constraining, though not fixing, the real description of the world in object-property language. If so, science would underdetermine reality, and there would be truths about the physical world invisible to the scientific method—perhaps permanently opaque to us, perhaps accessible through the methods of metaphysics. (The former position is one form of *epistemic structural realism* [Worrall 1989; Ladyman 1998], the view that science determines structure but not nature; the latter position characterizes the approach of metaphysicians of physics like Sider [2011; 2020], Maudlin [2018], and North [2021].) It is equally tempting to think that an interpretation of a formalized physical theory is a predicate precisification of a theory, telling us how uninterpreted mathematics is to be related to the concrete world.

Both temptations should be resisted. To make a mathematically stated theory subordinate to a natural-language description is to privilege natural language over other representational tools, and there is no non-question-begging reason to do so, and centuries of evidence from physics to the contrary (going back at least to Galileo's famous observation that mathematics is the language of nature). Language is an exceptionally good representational tool, but it is not our only such tool: Even prior to considering scientific theories themselves, think of maps or diagrams, of Kant's thesis of the centrality of the intuition of space, and of the manifold nonlinguistic forms of representation theorized in cognitive science.[13]

Nor can an "interpretation" be a predicate precisification, at least if understanding requires interpretation and physicists understand their own theories: Physics practice is not committed to any particular choice of precisification. What it is to learn a new physics theory, in practice, is not to be told how it maps to a world of objects and properties: It is to coordinate that theory with other theories, often mathematically presented, that are antecedently understood, to learn through those coordinatizations how the theory is to be used, and in due course to bootstrap one's way to an understanding of the theory that is not purely parasitic on previous theories. Much the same is true of learning a new language, or a new word in an old language, and if the process of doing so cannot be fully described except in terms of already-understood representations, that is just the familiar Neurath/Quine observation that we must rebuild our ship while sailing in it (Quine 1969, 126–127).

Opinions are much more finely grained than beliefs, but that fineness of grain does not necessarily reflect anything factive—sometimes it is just an artifact of the sharp edges that language requires.

13. See, e.g., Pitt (2022) and references therein.

This approach to scientific (or at least physics) theories might be called "mathematics-first realism." (It is a form of *ontic structural realism* [Ladyman 1998], though that term has been used in wildly different ways by different authors. I understand it broadly in the sense of McKenzie [2024].) Much more could be said in its explication and defense (I give my own version in much more detail in Wallace [2022; 2024]). For now, I will take as read that it is along the right lines and consider its implications for scientific ontology.

8.5 Ontology as Secondary

If the deep theories of physics do not represent the world in terms of objects and properties, does that mean that fundamentally there *are no* objects and properties? No, for two reasons. The first is shallowly semantic: To say that objects *do not exist* is already to concede the fundamentality of the framework for describing the world that does so in terms of objects, and then to advance a radical metaphysical thesis within that framework. ("There are no objects" is a claim in natural language.) The second is more substantive: Object-property talk is useful, even in deep physics. Giving accounts of a physical system in language is heuristically and pedagogically helpful; it builds intuition and helps coordinate different partial descriptions of the world with one another; it aids understanding and suggests new directions for work. A good predicate precisification of a theory is a valuable "tool for thinking," to borrow another term from Dennett (2013a). But it is not fundamental: Its claims are true, insofar as they are true, in virtue not of its correspondence with the world but of it being a predicate precisification of a representationally successful mathematized theory.

This makes the pursuit of ontology into a second-order activity: It tells us not directly what the world is like but how best we can describe it with the tools of natural language. The benefit of this somewhat deflationary account of ontology is that it liberates us methodologically. If we seek *fundamental truth*, then appeals to simplicity, to intuition, to fit with the manifest image, all seem question-begging: Why think that the world is simple, or that our intuitions are correct about it? But if we want that description in the language of a mathematized theory that is most helpful to us, then of course these are virtues: Who would choose a complicated, unintuitive predicate precisification when a simple, intuitive one is at hand? (It also permits agnosticism, even pluralism, when different, nominally incompatible predicate precisifications are tied for first place: Just use whichever one is more suitable for the task at hand.[14])

The approach to ontology I suggest here has much in common with Dennett's (2013) proposal for metaphysics as "sophisticated naïve anthropology," aimed at delivering a "metaphysics of the manifest image," without supposing that in doing so we are "limning the ultimate structure of reality." I am unsure quite how playfully Dennett intends his suggestion, but I intend mine entirely seriously: It is valuable to have clear predicate precisifications of theories in physics, and developing them is one (not the only) way philosophers of physics can contribute.[15]

And with this account of ontology in hand, we can finally sort out the metaphysical tangles that occupied us for the first half of the chapter. Consider once again our fluid, and for now assume (unrealistically) that we start off knowing only the lower-level theory, the mathematically presented theory of interacting atoms. That theory has a (very natural and obvious) predicate precisification: The objects are the atoms; their essential properties are their masses and perhaps charges; their spatial location can be understood as an additional primitive property or as a relation between particles and spacetime points (observe that even in this paradigm case we have underdetermination, invisible to physics).

That theory also, given certain auxiliary assumptions about dynamics and initial state, mathematically instantiates (approximately) the Navier-Stokes theory of fluid dynamics—that is, some subcollection of its degrees of freedom, appropriately coordinatized, can be well approximated as having self-contained and autonomous dynamics described by that equation. There is nothing pragmatic about this claim: It is simply a piece of mathematics, a theoretical discovery about the theory that we might be led to through surprising empirical data or just work out theoretically. And because of that piece of mathematics, among the scientifically interesting facts about the system are facts about those autonomous high-level dynamics, facts better studied via the Navier-Stokes equation than via interatomic physics. In studying those facts, it will often be helpful to have a description of fluid dynamics in object-property terms, and again there is a very natural and obvious choice of predicate precisification: The objects are the parts of the fluid, the properties include densities

14. Compare Feynman (1967, 162): "Every theoretical physicist who is any good knows six or seven different theoretical representations for exactly the same physics."

15. A few examples of philosophy of physics that fairly self-consciously fits that framework: Simon Saunders's (2013; 2016) work on individuality of particles; Eleanor Knox's (2013; 2014) spacetime functionalism (see also the discussion in Knox and Wallace [2023]); Chris Timpson and my spacetime state realism (Wallace and Timpson 2010; Wallace 2012, chap. 8).

(or perhaps masses) and temperatures, and parts of the fluid occupy parts of space.

What makes it true that, at atomic scales, there are atoms, is that we quantify over atoms in the predicate description we use for the system at those scales. What makes it true that at larger scales there are fluids and their parts is that we quantify over those parts in our large-scale description. And in each case our use of the description is legitimated by a combination of not-at-all-pragmatic facts about the mathematical representation of the system and somewhat-pragmatic facts about the best predicate precisification of that representation. Note that our ontology is *scale-relative* (Ladyman and Ross 2007, chap. 4): We quantify over fluids (when talking about large-scale physics) and we quantify over atoms (when talking about smaller-scale physics), but we don't quantify over both at once.[16]

This perspective on ontology also defuses the worry raised in section 8.3 about alien physicists with quite different pragmatic goals: Those aliens might disagree with us about *ontology*, which is a partially pragmatic matter, but (assuming both our and their science has been sufficiently successful) they will not disagree with us about mathematically expressed *structure*, which is not pragmatic at all. Sentient excitations of the magnetic fields in the stellar corona might not conceptualize particle physics in terms of object-like "electrons" and "protons"—indeed, the whole language-based idea of "object" and "property," grounded as it presumably is in the persistent and stable assemblies of matter that Earth's surface supports might be alien to them. But they can still agree with us about the mathematical structure of quantum electrodynamics and its representational accuracy to describe systems at appropriate length and energy scales.

What is the relation between the low-level and high-level ontology on this view? There need be no self-contained way of stating it, beyond the known relations between theory and ontology at each level and the equally known relation between theories.[17] Those relations, collectively, leave no metaphysical residue in need of explication. That said, there *might* be, in specific cases, and indeed the lack of any metaphysical requirement of a relationship gives us latitude to cash out those relationships in approximate or heuristic terms. There is clearly *some* sense in which a fluid is made up of atoms—the part of a fluid in region X has properties that supervene on the properties of the atoms

16. Not normally, anyway. Separation of scales is an important but not inviolate rule in physics: "The subatomic particle left a trail in the vapor in the cloud chamber" is an example of a scientifically legitimate use of mixed ontology.

17. Cf Guo (forthcoming), who contrasts this "theory-first" sense of reduction with the "ontology-first" approach presupposed in much metaphysics of science.

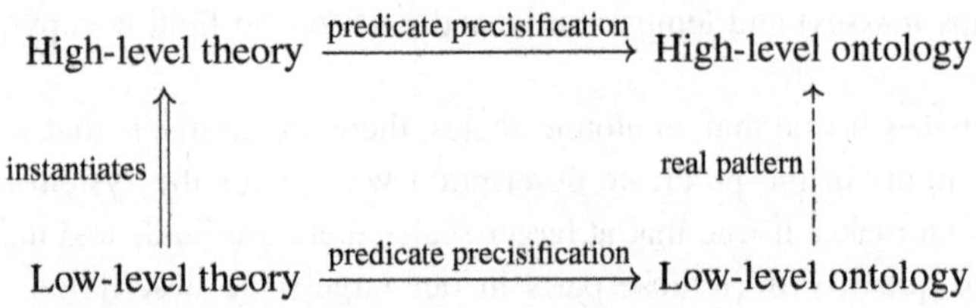

Figure 8.1: Theory and ontology in intertheoretic reduction.

in region *X*, and some of those properties (like mass) are simply additive functions of the properties. Is that a mereological relation . . . *sorta* (cf. Dennett 2013a, 96–97.) It's got a lot in common with the idea that mereology formalizes, for sure. Can we sharpen things up so that the relation is really-and-truly mereological, for some particular version of mereology? Maybe . . . but it's not entirely clear what would be gained in doing so.

Similarly, a hydrogen atom is *sorta* made up of a proton and an electron: There's a lower-level subatomic theory whose natural ontology is electrons and nuclei, a higher-level interatomic theory whose natural ontology is atoms, and a well-understood mathematical relation between the two, giving rise to a relation between the two ontologies that looks quite a lot like composition. A proton is *sorta* made up of quarks, but now there is a lot more heuristic and metaphor in that claim. In each case something like composition is happening, but the place to look if you want sharp and exact details is the mathematics, not the ontology, and when you look, you find the composition metaphor gets filled in and corrected very differently in different cases.

And in some cases composition does not even approximately capture the intertheoretic relation. Waves, and their quantized particles, are not even *sorta* made up of the medium that is waving: The right form of words is something closer to "moving disturbance." But still there will be some relation between lower and higher level ontology: Differences in high-level ontology reflect differences in high-level structure, which must result from differences in low-level structure, differences that are reflected in differences in low-level ontology. It would be good to have a general name for this relation, one that does not presuppose composition and reflects the structural, dynamical nature of the intertheoretic relations underlying it. I have a suggestion, one that I hope reflects the spirit of Dennett's own proposal: In this case, the higher-level ontology consists of *real patterns* in the lower-level ontology (figure 8.1).

8.6 Closing the Circle

Physics is atypical of the sciences: The systems it studies are more alien and unintuitive, and yet comparatively more simple, and its theories are mathematized to a far higher degree. The uncritical use of physics as a model for science more broadly has caused much confusion in philosophy of science. So I am hesitant to draw broader lessons about scientific metaphysics from my observations here. Yet my theme has been how ideas first developed in the philosophy of mind have found purchase in philosophy of physics, so it seems a shame not to *try* returning the favor. What follows should be taken in an appropriately tentative spirit, as an invitation to those who know the territory better.[18]

To begin with: Scientific theories outside physics are for the most part[19] not plausibly thought of purely as mathematical structures. Natural language runs much more deeply through these theories, far too deeply to be understood merely as a heuristic-but-dispensible predicate precisification.

I find this unsurprising, for two reasons. The first, and most obvious, is that most scientists study systems far more complicated than physicists do: No system in the institutional purview of physics has a fraction of the complexity of a bacterium, far less a human mind or an ecosystem or economy. To some degree, the mathematization of physics reflects the fact that precisely because physics is simpler, physicists have had much greater quantitative success in modeling the world.

But there is a deeper reason: Language as a representational tool evolved to help us get by on human scales, dealing with phenomena on scales of centimeters to kilometers, from milliseconds to decades. On those scales, the mismatch between even a physics description of the world and the description given in natural language is not that bad, especially if we are generous in how we construe "natural language" and allow it to include geometric concepts like angle or length, and certain abstracta like collections. It is only when we leave the regime of the macroscopic that linguistic descriptions show their limitations, and much of the special sciences is concerned with that macroscopic regime. So I would expect even a hypothesized "completed" biology or psychology

18. As an anonymous reviewer reminds me, theories of physics are also atypical in that the notion of *cause* plays a much more attenuated role in physics than in the rest of science. An account of scientific theories broad enough to cover (e.g.) evolution, or medicine, or economics, *clearly* would need to consider the role of cause (as recognized by Ladyman and Ross's [2007] extensive discussion of cause), but any such considerations lie beyond the scope of this article and I leave them again to those with greater expertise.

19. There are exceptions, for instance: some parts of economics, of quantum chemistry, of population genetics. Even here, though, the theory is usually to be seen as a mathematized chunk of some larger, more pluralistically formulated, representation.

(even supposing that makes sense) to be much more language-like than physics is.

But if most scientific theories are not collections of mathematical models, nor are they simply collections of natural-language sentences. I suspect the right way to describe them is a messy mixture of interconnected representational practices: Pieces of mathematics, chunks of natural language equipped with additional resources, concepts and tools and models borrowed from physics, and of course the large amount of tacit knowledge that one learns from experiment and observation and fieldwork and cannot losslessly describe. Much of how representation functions in this context will be at most partially and imperfectly described through the truth-satisfaction-reference. If so, supposing that "ontology" is the right fundamental tool to describe the world according to that theory may be as unwise as in physics, and we may do better, as in physics, to treat our fully linguistic descriptions of phenomena at least partially as a pedagogical and heuristic tool rather than the full and literal truth.

(Doing so, among other virtues, may help us to recognize that certain apparently legitimate questions arise because of the artificially sharp distinctions that language forces us to draw. I'll risk some tentative examples: "Is a fetus a human being?" "Are viruses alive?" and "Are wolves and dogs the same species?" are in each case grammatically well-posed questions, yet in each case it seems that we can have a perfectly satisfactory understanding of the underlying science without having an answer to that question.)

I'll finish by considering again the original question in the ontology of mind that provoked "Real Patterns" in the first place: On Dennett's account of intentionality, do people really have beliefs or is that just a useful fiction? Recall the context: We assume a person's behavior is pretty well, but very complicatedly, described by some comparatively low-level theory of the brain and body: Some "sub-personal cognitive psychology" (Dennett 1981).[20] But we also assume that same person can be quite well, and much less complicatedly, described, through the intentional stance: Through attributing to them beliefs and desires that are updated according to their informational environment and according to which their actions are rational. And the intentional stance description does not float free of the sub-personal explanation: It had better be the case that we can understand the lower-level mechanisms that make it the case that the intentional stance holds (and which in some circumstances, like brain damage or LSD, qualify or invalidate it).

20. A complication is that Dennett expects intentionality to recur, in simpler but robust form, at the sub-personal level; I put this aside for simplicity.

Does the person have beliefs? In the account of ontology I've offered (and throwing caution to the wind by applying it far outside physics), the answer is a clear YES: An ontology of beliefs and desires is the most helpful ontological description of the intentional-system theory, and that theory is correct, in the sense that it does indeed give a fairly robust, fairly reliable account of the person's behavior. To be is to be the value of a bound variable; our best predicate precisification of intentional systems theory quantifies over beliefs; so, at the scales and in the circumstances where we apply that theory, beliefs exist. Maybe it will turn out that those beliefs can be identified, exactly or approximately, with some sub-personal bit of ontology, but that's an additional scientific hypothesis not relevant to the higher-level ontology. The qualifiers on our ontological claim—"beliefs exist at certain scales and certain circumstances"—do not mark beliefs off as having some second-class status: All ontology works this way, and all higher-level ontological claims are claims about real patterns in lower-level ontology, and everything posited in physics outside the most extreme regimes of quantum gravity is "higher-level" in this sense. If beliefs don't *really* exist, then fluids don't *really* exist; atoms and electrons don't *really* exist; most likely space and time don't *really* exist.

There appears to be room for doubt about the reality of the beliefs and desires revealed by the intentional stance because there appears to be a contrast between abstracta, mere *patterns* in physical matter, and concrete things actually *comprising* of physical matter. But from the perspective of modern physics, this contrast is an illusion, caused by the fact that composition gives a reasonable—but far from perfect—account of intertheoretic relations at scales not too far from human imagination. Step back and look at the bigger picture, and it's real patterns all the way down.

Acknowledgments

In developing the ideas in this chapter I have benefited greatly from conversations with Neil Dewar, Alex Franklin, Bixin Guo, Eleanor Knox, James Ladyman, Kerry McKenzie, Tushar Menon, Katie Robertson, and Simon Saunders, and from discussions at the 2024 Santa Fe Institute workshop "Investigating Reality – A Philosophical, Mathematical, and Scientific Exploration." I am also grateful to Katie Robertson, Don Ross, and an anonymous reviewer for thoughtful comments on the manuscript.

I had hoped to use the opportunity of this chapter to thank Dan Dennett for the tremendous influence his work has been on my own, and for providing so compelling a model of how to do philosophy as a contribution to, and in communication with, science. As it is, I can only hope that in a small way it develops his ideas in ways that he would have approved of, in spirit if not in detail. *Ars longa, vita brevis.*

References

Albert, D. Z. 2000. *Time and Chance.* Harvard University Press.

Balescu, R. 1997. *Statistical Dynamics: Matter out of Equilibrium.* World Scientific.

Batterman, R. W. 2021. *A Middle Way: A Non-Fundamental Approach to Many-Body Physics.* Oxford University Press.

Cartwright, N. 1999. *The Dappled World: A Study of the Boundaries of Science.* Cambridge University Press.

Dennett, D. C. 1981. "Three Kinds of Intentional Psychology." In *Reduction, Time, and Reality,* edited by R. Healey, 37–62. Cambridge University Press.

Dennett, D. C. 1989. *The Intentional Stance.* MIT Press.

Dennett, D. C. 1991. "Real Patterns." *Journal of Philosophy* 88 (1): 27–51.

Dennett, D. C. 1995. *Darwin's Dangerous Idea: Evolution and the Meanings of Life.* Simon / Schuster.

Dennett, D. C. 2013a. *Intuition Pumps and Other Tools for Thinking.* W. W. Norton.

Dennett, D. C. 2013b. "Kinds of Things — Towards a Bestiary of the Manifest Image." In *Scientific Metaphysics,* edited by D. Ross, J. Ladyman, and H. Kincaid, 96–107. Oxford University Press.

Dupré, J. 1983. "The Disunity of Science." *Mind* XCII (367): 321–346.

Feynman, R. P. 1967. *The Character of Physical Law.* MIT Press.

Fodor, J. 1974. "Special Sciences (Or: The Disunity of Science as a Working Hypothesis)." *Synthese* 28:97–115.

Franklin, A. 2024. "Incoherent? No, Just Decoherent: How Quantum Many Worlds Emerge." *Philosophy of Science* 91:288–309.

Franklin, A., and K. Robertson. 2024. "Emerging into the Rainforest: Emergence and Special Science Ontology." *European Journal for Philosophy of Science* 14:61. http://philsci-archive.pitt.edu/19912/.

Frigg, R. 2007. "A Field Guide to Recent Work on the Foundations of Thermodynamics and Statistical Mechanics." In *The Ashgate Companion to the New Philosophy of Physics,* edited by D. Rickles, 99–196. Ashgate.

Guo, B. Forthcoming. "Ontology-First vs. Theory-First Approach to Reduction: A Case Study from Statistical Mechanics." *Philosophy of Science,* https://philsci-archive.pitt.edu/22733/.

Halmos, P. R. 1950. *Measure Theory.* Litton Educational Publishing.

Hermida, M., and J. Ladyman. 2022. *Living Objects.* https://philsci-archive.pitt.edu/id/eprint/21430.

Knox, E. 2013. "Effective Spacetime Geometry." *Studies in the History and Philosophy of Modern Physics* 44 (3): 346–356.

Knox, E. 2014. "Newtonian Spacetime Structure in Light of the Equivalence Principle." *British Journal for the Philosophy of Science* 65 (4): 863–880.

Knox, E., and D. Wallace. 2023. "Functionalism Fit for Physics," https://philsci-archive.pitt.edu/22655/.

Ladyman, J. 1998. "What Is Structural Realism?" *Studies in History and Philosophy of Science* 29 (3): 409–424.

Ladyman, J., and D. Ross. 2007. *Every Thing Must Go.* Oxford University Press.

Lewis, D. 1970. "How to Define Theoretical Terms." Reprinted in David Lewis, *Philosophical Papers*, Volume I (Oxford University Press, Oxford, 1983), *Journal of Philosophy* 67 (13): 427–446.

Maudlin, T. 2018. "Ontological Clarity via Canonical Presentation: Electromagnetism and the Aharonov–Bohm Effect." *Entropy* 20 (6): 465.

McKenzie, K. 2024. "Structuralism as a Stance." *Philosophy of Physics* 2 (1): 1–23.

Myrvold, W. 2022. "Philosophical Issues in Thermal Physics." *Oxford Research Encyclopedia in Physics,* https://doi.org/10.1093/acrefore/9780190871994.013.42.

Nagel, E. 1961. *The Structure of Science.* Harcourt.

North, J. 2021. *Physics, Structure, and Reality.* Oxford University Press.

Pitt, D. 2022. "Mental Representation." In *The Stanford Encyclopedia of Philosophy,* Fall 2022, edited by E. N. Zalta and U. Nodelman. Metaphysics Research Lab, Stanford University.

Price, H. 1996. *Time's Arrow and Archimedes' Point.* Oxford University Press.

Quine, W.V.O. 1948. "On What There Is." *Review of Metaphysics* 2 (5): 21–38.

Quine, W.V.O. 1969. "Natural Kinds." In *Ontological Relativity and Other Essays,* 114–138. Columbia University Press.

Ross, D. 2000. "Rainforest Realism: A Dennettian Theory of Existence." In *Dennett's Philosophy: A Comprehensive Assessment,* edited by D. Ross, A. Brook, and D. Thompson, 147–168. MIT Press/Bradford.

Saunders, S. 2013. "Indistinguishability." In *The Oxford Handbook of Philosophy of Physics,* edited by R. Batterman, 340–380. Oxford University Press.

Saunders, S. 2016. "The Emergence of Individuals in Physics." In *Individuals Across the Sciences,* edited by A. Guay and T. Pradeu, 165–192. Oxford University Press.

Seifert, V. A. 2023. "The Chemical Bond Is a Real Pattern." *Philosophy of Science* 90 (2): 269–287.

Sider, T. 2011. *Writing the Book of the World.* Oxford University Press.

Sider, T. 2020. *The Tools of Metaphysics and the Metaphysics of Science.* Oxford University Press.

Sider, T. 2024. *3D in High-D.* https://tedsider.org/papers/3D_in_high-D.pdf.

Sklar, L. 1993. *Physics and Chance: Philosophical Issues in the Foundations of Statistical Mechanics.* Cambridge University Press.

Wallace, D. 2003. "Everett and Structure." *Studies in History and Philosophy of Science Part B: Studies in History and Philosophy of Modern Physics* 34 (1): 87–105.

Wallace, D. 2010. "Decoherence and Ontology: Or: How I Learned to Stop Worrying and Love FAPP." In *Many Worlds? Everett, Quantum Theory, and Reality,* edited by S. Saunders, J. Barrett, A. Kent, and D. Wallace, 53–72. Oxford University Press.

Wallace, D. 2012. *The Emergent Multiverse: Quantum Theory According to the Everett Interpretation.* Oxford University Press.

Wallace, D. 2013. "The Arrow of Time in Physics." In *A Companion to the Philosophy of Time,* edited by H. Dyke and A. Bardon. John Wiley / Sons.

Wallace, D. 2016. "Probability and Irreversibility in Modern Statistical Mechanics: Classical and Quantum." In *Quantum Foundations of Statistical Mechanics,* edited by D. Bedingham, O. Maroney, and C. Timpson. Oxford University Press, forthcoming. https://arxiv.org/abs/2104.11223.

Wallace, D. 2022. "Stating Structural Realism: Mathematics-First Approaches to Physics and Metaphysics." In *Philosophical Perspectives Volume 36: Metaphysics,* edited by J. Hawthorne, 345–378. Wiley-Blackwell.

Wallace, D. 2023. "Thermodynamics With and Without Irreversibility." In *The Arrow of Time: From Local Systems to the Whole Universe,* edited by O. Lombardi and C. Lopez. Cambridge University Press, forthcoming. https://philsci-archive.pitt.edu/22273/.

Wallace, D. 2024. *Learning to Represent: Mathematics-First Accounts of Representation and Their Relation to Natural Language.* https : //philsci-archive.pitt.edu /23224.

Wallace, D., and C. Timpson. 2010. "Quantum Mechanics on Spacetime I: Spacetime State Realism." *British Journal for the Philosophy of Science* 61 (4): 697–727.

Worrall, J. 1989. "Structural Realism: The Best of Both Worlds?" *Dialectica* 43 (1-2): 99–124.

9 Disinterested Taxonomic Pluralism: Biological Classifications Track Real Patterns

Margarida Hermida

9.1 Introduction

Two main views of scientific classification are often contrasted. On the one hand, there is a realist or naturalist view, according to which, when scientific classifications get things right, they correspond to real divisions that exist in nature, independently of the minds of the classifiers. This view contrasts with a conventionalist or constructivist view that holds that any classification always reflects the interests of the classifiers, with scientists' interests being merely one class of interests among many (Bird and Tobin 2024). It is natural to think, however, that if there are real divisions in nature, then there must be one single correct way of classifying things; and in contrast, when we find multiple classifications within a particular scientific domain, this must mean that they are conventional, or serve different interests of the classifiers. Thus, realism tends to be associated with monism, and taxonomic pluralism with interest-relativity. Scientific classification if often presented in terms of this dichotomy:

> Are there objective, real divisions in nature, independent of the mind, on which a single correct way of classifying things eventually discovered by science would be grounded? Or, on the contrary, do scientific classifications always reflect to some extent the aims and interests of the classifiers? (Ruphy 2016, 110–111)

Despite its intuitive appeal, I want to argue that it is a false dichotomy, as it conflates two questions: (1) whether there is one single way of classifying things within each domain of science, or more than one; and (2) whether, given pluralism, different classifications reflect real divisions in nature, or merely reflect the interest of the classifiers. The existence of more than one classification within a particular domain does not automatically imply that classification is interest-based, subjective, or arbitrary. Rather, it might just mean that there is more than one set of real divisions in nature. Naturalness does not imply uniqueness.[1]

Biological classification is a source of examples of pluralism and, because this distinction is not made, it is also used to support arguments in favor of interest-relativity. Thus, while monists point to "the stability and epistemic fecundity of classifications in physics and chemistry" (Ruphy 2016, 111), pluralists often "appeal to the plurality of classifications in biology to make a case for a lack of natural order and for the dependency of the classifications on the aims and interests of the classifiers" (111). If there are indeed multiple classifications in biology, *and* if this plurality is due to the different interests of the classifiers, then that is a problem for scientific realism. The worry is that, if classifications depend on the aims and interests of the classifiers, they might be subjective to some extent and, therefore, fail to adequately describe the mind-independent world that science aims to investigate.

Dupré's promiscuous realism is an attempt to square the circle, by recommending realism about *all* the interest-based classifications—including nonscientific ones. In doing so, however, it quickly drifts away from scientific realism and into disputes about ordinary language, as when Dupré claims that "What a fish is is not the sort of thing a scientist (except, perhaps, a linguist) could find out" (1999, 467). The meaning of the common term "fish" (which might include such disparate items as "jellyfish" and "shellfish") has nothing to do with scientific classification, and is surely irrelevant for determining where to place one's ontological commitments. On the contrary, if, as Ereshefsky (2022) notes, "species" is a theoretical term from evolutionary biology, then its meaning outside of biology does not count, any more than we need to consult ordinary language meanings of the term "work" to understand the theoretical physical concept; and the same can be said of the names of particular biological taxa such as "Chondrichthyes" and "Actinopterygii."

Ruphy suggests that the impasse between monism and pluralism can be resolved by embracing scientific disunity, for example by stating that biology is fundamentally different from physics and chemistry when it comes to classification (Dupré 1993, 84). This solution, however, involves accepting not only a disunified view of science but also the disunified metaphysical view of the world that goes with it—the "dappled" world that is orderly in some domains but not others. Alternatively, the impasse can also be overcome if one of the views is clearly better than the other. Thus, monists could reject the idea that

1. Franklin-Hall also points out that, although realists tend to be categorical monists, realism is compatible with the existence of more than one classification, including cross-cutting ones (2015, 929).

taxonomic pluralism in biology has any metaphysical consequences by alleging that it stems from the current immaturity of biology in comparison with the physical sciences; or pluralists might argue that, despite appearances, classifications in physics and chemistry are also dependent on the interests of the classifiers (Ruphy 2016, 111–112).

There is a better way. In this paper, I propose to overcome the false dichotomy of one real objective way of classifying things vs. many interest-based classifications by appealing to Dennett's (1991) notion of *real patterns*. Within any domain of science, things can often be classified in more than one way because there are multiple real divisions in nature, which correspond to multiple real patterns. Although others have argued for the validity of cross-cutting scientific classifications (e.g., Khalidi 1993, 1998, 2013; Tobin 2010; Havstad 2021), they usually justify them in terms of the different epistemic interests of the classifiers. For example, Khalidi states that the existence of crosscutting classifications in biology can be explained "by saying that they embody different *interests* relative to a common subject matter," (1998, 42), and Tobin states that:

> Pluralism is the view that there are several different equally legitimate ways of carving nature into natural kinds. For example, in biology pluralists do not believe that there is a single species concept. They argue that there are a number of legitimate species concepts, which divide species *according to different interests*. (Tobin 2010, 189, n. 3, emphasis added)[2]

While some cross-cutting classifications may indeed reflect different epistemic interests, this is not necessarily the case for all cross-cutting classifications, and indeed it is not the case for some of the most frequently discussed cases of biological pluralism. Dennett's real patterns theory provides a framework that illuminates the nature of scientific classification, and allows for exactly as much pluralism as is required to capture the multiple real divisions in nature that track different kinds of phenomena—and no more. Pluralism is not about "letting a thousand flowers bloom" or having indefinitely many classifications, but it does mean often having more than one, or a few, natural classifications within a particular domain, where each natural classification tracks one

2. Similar views are found in the context of biomedical science. For example, Anya Plutynski (2018) notes that "different agents with different purposes (surgeons, oncologists, pathologists, epidemiologists) have different epistemic (and practical) interests, which yield distinct sorting principles and distinct classifications of tumor types and subtypes" (2018, 51). In this chapter, I abstain from discussing cases from biomedicine, where interests might plausibly play a more prominent role. However, Khalidi (2024) argues that, even in applied areas, non-epistemic interests and values should not play a role in determining the classifications themselves, but only in deciding which one to apply in a particular case.

real pattern discovered by science. The outline of the chapter is as follows. The following three sections discuss cases of taxonomic pluralism in biology: Section 9.2 focuses on pluralism about species concepts, section 9.3 discusses the three main taxonomic schools in the twentieth century—evolutionary taxonomy, phenetics, and phylogenetics—and section 9.4 focuses on pluralism involving phylogenetics and the various non-genealogical classifications that are used alongside it, such as ecological classifications. Section 9.5 argues that natural classifications capture real patterns whether or not they involve natural kinds, and shows how the real patterns framework can be applied to various components of scientific ontology. Section 9.6 is a brief conclusion.

9.2 Species Concepts

The variety of species concepts is the most discussed case of taxonomic pluralism in biology, but there is a lot of hype about it. There are indeed multiple species concepts; Zachos (2016) compiled a list of thirty-two. However, many of these are not distinct concepts, but variations on only a few main concepts. In addition, some concepts are either not in use anymore, or are only used as a *proxy* or *surrogate* for a more adequate species concept when the evidence at hand does not allow, or practical difficulties impede, its use (e.g., Derraik et al. 2010). For example, *morphospecies* are sometimes used in paleontology when only morphological data is available, even though many paleobiologists think that most fossil morphospecies actually correspond to species groups or genera (Allmon 2016). This is not a case where "different scientific interests and values lead to different but equally valid ontologies" (Ludwig and Ruphy 2021), as the biologists who use these operational concepts in epistemically un favorable situations acknowledge their inadequacy and do not claim that they capture the real *biospecies*.

In a recent study, Stankowski and Ravinet (2021) helpfully quantify how much disagreement there is currently in biological taxonomy, by surveying practicing biologists working in different disciplines, model organisms, and problems. They remark that "in terms of the number of concepts used, the disagreement is not as extensive as one might have expected" (Stankowski and Ravinet 2021, R429). In fact, most researchers (around 90%) use one of only four concepts: The *biological species concept* (BSC), which recognizes species based on reproductive isolation; the *evolutionary species concept* (EvSP), according to which species are cohesive evolutionary units; the *genotypic cluster species concept* (GCSC), which states that a species is a genetically distinguishable group of individuals that has few or no intermediates when in contact with other such clusters, and the *phylogenetic species concept* (PSC), according to which a species is the smallest monophyletic group of common ancestry (Simpson 1961; de Queiroz and Donoghue 1988; Mallet 1995; Mayr 1995; Stankowski and Ravinet 2021).[3] Evidently, the amount of disagreement is overestimated when slight variants of the same concept, which differ, for instance, in *exactly how much* reproductive isolation is needed to recognize species, are taken to be distinct concepts.

Even if most biologists use only one of four main concepts, that still looks like a remarkable amount of disagreement. Stankowski and Ravinet's study offers some clues toward explaining this pluralism: Among the factors they analyzed, the most significant effects were the *evolutionary scale* of the research, and the subdiscipline. Researchers who study *microevolution*—that is, population-level change over small time scales, as well as the speciation process itself—are much more likely to use the BSC; whereas those who study *macroevolution*, involving change across deep time and higher taxonomic groups, were more likely to use the EvSC or the PSC. The same pattern was broadly observed for more ecologically focused disciplines vs. more historically focused ones (Stankowski and Ravinet 2021).

Surprisingly, the authors conclude that "concept use is shaped by various factors that are essentially cultural differences between groups of biologists," and that pluralism concerning species concepts reflects the "extremely diverse" goals of biologists (Stankowski and Ravinet 2021, R429). This seems to be in line with interest-relativity accounts of pluralism. But it is entirely the wrong conclusion to draw from their study. What their results indicate instead is that researchers working on different evolutionary timescales tend to use different species concepts, which can be explained by the need to pick out the relevant pattern that is prevalent at that scale.

Let's start by examining de Queiroz's proposed solution to the species problem. According to de Queiroz, it is possible to unify all species concepts into a more general concept by finding a common element. He suggests that the common element is that species are "separately evolving metapopulation lineages" or, more specifically, segments of such lineages (2007, 880–881). The different species concepts then differ only in what exactly separates those metapopulation lineages: Reproductive isolation, monophyly, diagnosable phenotypic differences, and so on. The crucial point is that, since these do not arise at the same time during the protracted process of speciation, there is scope for disagreement as to which one is more important—which corresponds to disagreement over *at which point* in the speciation process there are distinct

3. Convergence among these concepts may be even greater. For example, some authors consider the BSC to be a variant of, or coincidental with, the EvSC, at least for biparental species (Frost and Kluge 1994; Wiley and Lieberman 2011), and genotypic clustering can be used as evidence for the delimitation of species on the EvSC (Barraclough 2019).

species. Zink and Klicka (2018) describe how, during the process of speciation, two diverging lineages might be recognized as evolutionary species under the EvSC as soon as they diverge and gene flow ceases; later on, "genetic drift, natural selection, or sexual selection will yield diagnostic traits that, under the PSC, allow members of each lineage to be identified as distinct phylogenetic species" (51). Still later, reproductive isolation mechanisms will have evolved, which then allow the two lineages to be recognized as biological species under the BSC.

On this view, each species concept identifies one specific pattern in the relations between populations during the speciation process: (1) cessation of gene flow, (2) emergence of recognisable differences between the two populations, and finally (3) the establishment of reproductive isolation mechanisms. Nevertheless, even if de Queiroz is right about the common element of all species concepts and where the differences come from, recognising that different species concepts apply to species at different times during the extended process of speciation does not necessarily lead to consensus. For example, proponents of the BSC can simply claim that evolutionary and phylogenetic "species" are mere varieties or incipient species, on their way to becoming, but not yet "good and distinct species" (Darwin 1859, 61), because they are not reproductively isolated—and indeed, the speciation process might even be aborted if, for example, barriers to gene flow are removed before it is complete. Thus, it is unlikely that pluralism will be completely eliminated. Each species concept captures an important component of speciation that plays a role in the evolutionary process, and is a real pattern in the world that is recognized by biological theory.

This is helpful, but it still doesn't capture the whole story. In order to understand species concept pluralism, it's important to take into account not only processes that happen at different times during speciation but patterns at vastly different temporal and evolutionary scales. At the microevolutionary scale, population-level effects dominate, and speciation is a long and complex process that might start with the appearance of a physical barrier and only be complete thousands of years later with the establishment of reproductive isolation mechanisms. At the macroevolutionary scale, however, population-level processes are invisible, speciation can be approximated as instantaneous, and the separately evolving lineages (and the clades they give rise to) are the only relevant patterns.

Switching to a different level or scale often involves an ontological shift (Dennett 1991, 39). That this is clearly a metaphysical, as opposed to a merely epistemic shift, was most clearly formulated by Ladyman and Ross in their thesis of the *scale relativity of ontology*, which states that "what (really, mind-independently) exists should be relativized to (real, mind-independent) scales at which nature is measurable" (2007, 200). Different patterns exist at different scales. In the case of biological species, *interbreeding populations* exist at demographic scales, whereas *evolving lineages* only exist on much longer evolutionary time scales.

It is not surprising, then, that macroevolutionary researchers tend to use evolutionary and phylogenetic species concepts, whereas ecology, population biology, and the study of microevolution, including speciation, calls for the biological species concept. Okasha (2002, 201) notes that "phylogenetic [and evolutionary] species concepts are intended to apply over evolutionary time, while the other [concepts] are meant to apply to contemporaneous organisms." The fact that evolutionary and phylogenetic concepts refer primarily to *lineages* as opposed to *populations* or *individuals* is a clue to their temporal scope. However, even temporally extended concepts such as the phylogenetic species concept have to rely on a contemporaneous concept in order to account for speciation events—that is, for the splitting of one lineage into two (Okasha 2002).

Different species concepts—insofar as they are actually different—track the real patterns of speciation and separately evolving lineages at different evolutionary and temporal scales. Taxonomic pluralism about species concepts is a direct consequence of the existence of more than one real pattern in the process of speciation and evolution of species—namely, patterns that are found at different temporal scales.

9.3 Taxonomic Schools

A second case of pluralism in biology concerns the dispute between three main schools of taxonomy in the twentieth century: Evolutionary taxonomy, phenetics or numerical taxonomy, and phylogenetics. Although it is a case of temporary disagreement rather than persistent pluralism—phylogenetics eventually became the accepted system of classification—real patterns are relevant to the explanation of the failure, and success, of the different classification systems.

Darwin noted that many naturalists thought that "something more is included in [biological] classification, than mere resemblance," and he suggested that "propinquity of descent" was the bond, "hidden as it is by various degrees of modification, which is partly revealed to us by our classifications" (1859, 413–414). Hidden behind the hierarchy inherent in the Linnaean taxonomy was a system of genealogical relations. In the twentieth century, *evolutionary taxonomy* attempted to produce a properly genealogical, evolutionary classification (Richards 2011, 101–119). In this system, biological species were grouped together in higher taxa in virtue of their common ancestry, which was determined on the basis of shared homologies. Higher taxa were supposed to be "monophyletic" in the sense that they contained only the descendants of a single ancestral species (Mayr 1942, 280; Richards 2011, 105).

However, Mayr's notion of "monophyly" allowed for a taxon to be paraphyletic—that is, to include only descendants of a single ancestral species, but not all of them. Thus, for example, in evolutionary taxonomy, the birds and the mammals were placed in separate taxa and given a rank similar to that of the reptiles, because both groups exhibit a vast new array of characters and great diversification. This makes Reptilia a paraphyletic taxon, because it did not include all the descendants of a single stem species. This introduced a subjective component into the classification. What is sufficiently novel, or what counts as a suitable degree of diversification, were questions left to the judgment of the taxonomist.

Even more problematic was the placement of species into genera, which was partly based on degree of divergence but also on practical considerations. The genus was considered "a convenient category" in taxonomy (Mayr 1942, 283), which should be kept neither too large nor too small (Richards 2011, 107). If a genus was excessively speciose, a systematist would be justified in splitting it. Again, it was the opinion of the systematist that counted. Thus the classification reflected phylogeny to some extent, but it also served the practical function of "breaking up groups of species into manageable chunks," and was allowed to be determined to some extent by the individual preferences of the systematist (Richards 2016, 108). These subjective elements made evolutionary taxonomy a partly natural, partly conventional and *interest-based* classification—a factor which ultimately contributed to its rejection.

Another important school of classification in the twentieth century that has now been abandoned was numerical taxonomy, also known as phenetics (Sokal and Sneath 1963; Richards 2016). Pheneticists took the radical approach of excluding evolution from classification, which they thought made the classification somewhat circular (since phylogeny was inferred on the basis of shared homologies, but homology was also inferred on the basis of evolutionary relationships).[4] Instead, they suggested that classification should be

4. See Hull (1967) for a rebuttal of the charge of circularity in evolutionary taxonomy.

based on measurable similarities and differences among organisms—*all* similarities and differences, with no distinction made between homologies and homoplasies.

Although the numerical taxonomists were intent on getting rid of what they thought was an unacceptable degree of subjectivity in classifying traits as homologies or homoplasies, their own system was riddled with subjective elements: Firstly, deciding what counts as a character is not straightforward—organisms do not come neatly divided into "units of information" with only one way of being logically subdivided. Then, there is the problem of coding. For example, Sokal and Sneath (1963, 70) suggest that plant leaf length could be coded as "short" or "long." But surely it could also be coded as "short," "medium," and "long," or any number of arbitrary subdivisions of length (Richards 2016, 123). Determining how similar a character must be for it to count as the same one in different organisms can also be subjective. Finally, once all characters are selected and coded, there are still a number of different algorithms that could be used to determine a coefficient of similarity based on all that information, and no principle to guide its selection. Partly due to these problems, phenetics was never widely accepted.

The pheneticists were right, however, in one particular criticism of evolutionary taxonomy, namely that, even if the true phylogeny was known, a single classification "could not simultaneously represent all the important features of evolutionary relationships" (Richards 2016, 120). Indeed, evolutionary taxonomy aimed to represent "genealogy, evolutionary diversification, and morphological divergence" (144). But the fact that the system tried to represent multiple patterns at once meant that it was unable to represent any of them adequately, due to the need to simultaneously represent the others. For example, the separation of the birds and mammals from the reptiles may have adequately represented the high degree of evolutionary novelty and diversification of those groups, but it also meant that phylogenetic history was not accurately represented.

In contrast, phylogenetics aims to represent only the branching structure of evolutionary history (i.e., the relations of ancestry and descent among lineages of organisms). In phylogenetic systematics, only strictly monophyletic groups, or clades, are accepted. A clade must include all, and only, the descendants of an ancestral species. Since the groups are determined by the actual branching pattern, there is only one possible way to group taxa, thus eliminating the unacceptable subjectivity of evolutionary taxonomy: In phylogenetics, "the judgment of different systematists cannot produce different groupings"

(Richards 2016, 127).[5] This makes the classification—insofar as it gets the real phylogenetic relationships right—entirely natural. While paraphyletic groups are "created by systematists, not by nature" (Wiley and Lieberman 2011, 19), monophyletic groups are a real feature of the natural world that exists in virtue of how the evolutionary process operates. Their existence is, in fact, predicted by evolutionary theory: "Evolutionary theory predicts that monophyletic groups and only such groups emerge from various evolutionary processes termed speciation" (Wiley and Lieberman 2011, 18).

As noted above, the pluralism involved in the dispute between different schools of taxonomy was temporary and was eventually resolved by the rejection of evolutionary taxonomy and phenetics in favor of phylogenetic systematics. There may be a case for saying that, in this instance, pluralism did stem from a certain immaturity of the science (as Ruphy suggests monists might argue). But the reasons for the rejection of the other taxonomic schools are instructive. It is far from the case that biologists are satisfied to countenance multiple interest-based classifications. On the contrary, both phenetics and evolutionary taxonomy involved a degree of subjectivity that was unacceptable to most biologists, and this contributed to their rejection. Furthermore, evolutionary taxonomy tried to capture multiple patterns simultaneously—genealogy, diversification, and morphological divergence—with the result that it failed to capture any of them adequately. Phylogenetics, on the other hand, restricted itself to capturing exactly one pattern in the evolutionary history of all organisms on Earth—their relationships of ancestry and descent—and this was key to its success.

9.4 Phylogenetics and Other Classifications

For someone engaged in advocating for the advantages of phylogenetic systematics over other classification systems, Hennig (1966) is remarkably pluralist:

> [Organisms] in their totality form a "multidimensional multiplicity" . . . Therefore it is possible to arrange animated natural things in numerous different systems, depending on which of . . . different relationships has been investigated. The differences among all these systems are determined by the particular relationships of which they are a concrete expression. (Hennig 1966, 4)

5. Disagreement may still occur in the use of different computational methods of phylogenetic inference, such as maximum parsimony, Bayesian inference, or maximum likelihood. Phylogenetic trees produced by scientists represent hypotheses concerning the actual relationships among organisms, given the available evidence, with a certain degree of confidence. But assuming the actual relationships are correctly inferred from the data, there is only one way to group taxa.

What's more, these different systems, if scientifically sound, are equally valid, as long as they capture actual relations among natural objects. There is no scope for subjectivity or arbitrariness; each system earns its place in science because it captures one dimension of the "multiplicity":

> All these different systems are, fundamentally, equally justified so long as they are a proper expression of the membership position that an object of nature possesses within the framework of the totality, for the dimension that was chosen as the basis for the particular system. On the other hand, every system is fully and unequivocally determined by the kind of relations within the multidimensional multiplicity of which it is an expression; or to express it another way, by the structure of that particular totality of which the individual things should be understood to be members, so that there cannot be any arbitrariness in the formulation of the system. (Hennig 1966, 4)

This talk of positions of objects in a multidimensional multiplicity can more perspicuously be translated into the concept of real patterns in the structure of the world. According to Dennett, there is a real pattern in some data if there is a description of the data that is more efficient than the bitmap (1991, 34). Natural classifications capture real patterns because they provide more efficient descriptions of the data than a shopping list of all the individual things they classify, and the reason they are able to do that is because they identify projectible regularities in nature. There are, however, multiple real patterns in the world, often even within a particular domain, and each classification that captures one such pattern—one "dimension of the multiplicity"—is a natural classification, despite the fact that the various classifications cross-classify the same entities. In fact, we should expect them to cross-classify, as the various patterns can be orthogonal to each other. Hennig points to ecological (e.g., *parasite*, *predator*, *saprophyte*), physiological (e.g., *ectotherm* vs. *homeotherm*), and biogeographical classifications as examples of (often partial, and not always systematized) classifications "carried out for different dimensions of the multidimensional multiplicity" (1966, 5).

Here are a few more examples. Khalidi (1993) notes that developmental stages of insects (*larva*, *pupa*, *imago*) cut across taxonomic categories; so do more general developmental stages of animals, such as *blastula* and *gastrula*. Metabolic kinds, such as *phototroph*, *chemotroph*, and *heterotroph*, classify organisms in terms of the source of energy utilized. Organisms may also be classified in terms of hierarchical structure, as *unicellular*, *multicellular*, and *colonial* organisms. In the reproductive patterns of animals with sexual reproduction one can distinguish between *oviparity* and *viviparity*, and the mode of provisioning of the developing embryo can be classified as *lecithotrophic* or *matrotrophic*, of which there are further sub-kinds.

Phylogenetic classification captures a real pattern—the structure of historical relations of ancestry and descent among all the organisms that descend from a single origin of life event. But other classifications, such as ecological, metabolic, or functional ones, capture other patterns that are no less real. For example, there are a limited number of ways in which living organisms can obtain free energy from the environment to power their activities—a metabolic classification can capture these regularities. Ecological classifications capture the regularities involved in the interactions of organisms with other organisms and with the abiotic environment. And so on.

It is natural to think that a phylogenetic classification might be more *useful* for evolutionary biologists, or that ecologists might prefer to use an ecological classification. There is the idea that "ecologists do not group [organisms] according to the same parameters that [other] biologists do" (Ruphy 2016, 126), and that due to their different epistemic interests, "ecologists classify the domain of organisms into groups such as predator and prey, while evolutionary biologists classify the same domain into genealogically determined species groupings" (Franklin-Hall 2015, 933). But this is not exactly the case. Species are the main unit for ecological models of diversity (Barraclough 2019, 13), and phylogenetics serves as a reference classification system for all biological studies, including in ecology. While population ecologists may use other classifications *in addition* to phylogenetics, an ecology study can hardly get off the ground without identifying the organisms under study, which means locating them on the phylogenetic tree. For example, although parasitology is focused on a particular type of ecological relation among organisms (Khalidi 1998, 42), this does not preclude its use of phylogenetic systematics in the classification of both parasites and their hosts. On the contrary, in order to determine the specificity of a parasite, the parasitologist needs to inquire about the species of hosts it occurs in; and in order to determine important ecological parameters of parasite communities, such as species richness, diversity, and dominance, they need to identify each parasite to the lowest possible taxon.

Studying the exact same phenomenon often requires more than one classification system. Evolutionary biology is a good example. Distinguishing whether two organisms share a particular feature because they are closely related or because they have independently converged on the same trait due to similar selective pressures is a crucial question in evolutionary biology (McGhee 2011, 2), and for this, evolutionary biologists need both phylogenetic groups and functional kinds. Equally, a complete understanding of the various patterns in the functioning of living organisms requires an understanding of how they emerge in the context of evolutionary theory. So it is not the case that different classifications are motivated by different interests within a particular domain; different classifications may be required even for the study of one particular research question within the same domain.

This observation also counters a different kind of monism, one that is motivated by interest-relativity, namely, the categorical bottleneck account of natural kinds—that is, the idea that natural categories are those that serve the interests of a wide variety of epistemic agents (Franklin-Hall 2015). Although the fact that a particular classification is widely endorsed and useful to many epistemic agents may be defeasible evidence of its naturalness, the restricted "interest" of a classification—as in the example of a morphological classification that captures patterns of biomechanical features involved in animal locomotion, but is overall less useful for many purposes than an evolutionary classification (Franklin-Hall 2015, 942)—does not imply it is unnatural.

Even if some classifications are more useful in certain subfields than others, interest-relativity still does not explain the existence of the different classifications in the first place. Rather, "researchers have incentive to generate any and all those classification systems which grant them reliable, fruitful inferences" (Havstad 2021), and because there are multiple patterns in the properties and relations of complex objects in the world, this inevitably leads to multiple scientific classifications, each of which captures one particular pattern. Even if "Patterns of inference bearing on certain themes . . . do tend to be differentially valuable from different perspectives," Havstad argues, "this interest-relativity is not what generates or explains the naturally divergent kinds; rather, it motivates and rewards our recognition of them" (2021). The next section considers the importance of real patterns for scientific classification and for science more broadly.

9.5 Real Patterns and Scientific Classification

Philosophical discussions of scientific classification tend to focus on the issue of natural kinds. In philosophy of biology, an important insight was the recognition that species cannot be accommodated within traditional essentialist accounts of natural kinds, and should instead be conceived of as particulars—complex historical individuals, with a beginning (speciation) and an end (extinction), and whose properties can change over time as a species evolves.[6]

6. They can be considered natural kinds on some accounts, with the caveat that they are evolving kinds, with no essential intrinsic properties—although they may have essential relational properties (Okasha 2002). However, organisms within a species can differ dramatically in their properties, for example in cases of extreme sexual dimorphism, polymorphism, and alternation of generations. Partly for this reason, although HPC kinds are a popular strategy among philosophers to accommodate biological taxa within the natural kinds framework, they arguably conflict with biological practice (Ereshefsky and Reydon 2015).

But if species and higher taxa are not natural kinds, this does not make them any less real, or phylogenetic classification any less natural. It is important to distinguish the question of kindhood from the question of naturalness—a classification of organisms into phylogenetic groups can be natural, even if these groups are not kinds. Realism about biological species and higher taxa does not require treating them either as natural kinds or as individuals. Instead, one can simply be a realist about the pattern—the structure of relations among organisms. Species and higher taxa are real patterns in the structure of relations among individual organisms in their environments, and over evolutionary time. Consider Hennig's comments on the reality of human families, which applies equally well to phylogenetic groups:

> There are still many families that were torn apart by the war. The members of these families know nothing of each other. Children have grown up in the meantime. No one doubts that such families exist. Their boundaries as "supra-individual groups" are determined by the tokogenetic relationships that exist between the members of the families. (Hennig 1966, 80)

Equally, natural taxa, including species and monophyletic groups, "exist whether or not there are any taxonomists around to perceive and name them" (Wiley and Lieberman 2011, 68), regardless of whether they are kinds. Changing the focus from natural kinds to patterns provides a more solid basis for scientific realism. Scientific realists want to be realists about scientific ontology. But scientific ontology can include *things* (e.g., material objects, such as organisms, cells, ribosomes, viruses, and atoms), *processes* (e.g., photosynthesis, cell division, and volcanic eruptions), and also *relations* or *interactions* between things (e.g., the hydrogen bonds between two strands of DNA, phylogenetic relations among species, and interactions between organisms in ecosystems). All of these various components of scientific ontology can be captured in terms of real patterns.

Natural kinds are particularly good candidates for being real patterns, since one of their aims is precisely to encode projectible regularities in a particular domain. Thus, Ladyman and Ross propose that a natural kind should be understood as merely "a real pattern with high indexical redundancy"—that is, a real pattern that is measurable from many locators (2007, 205), and argue that "everything a naturalist could legitimately want from the concept of a natural kind can be had simply by reference to real patterns" (294). Crețu also thinks that natural kinds should be conceived of as real patterns, understood as "robust relations among entities exhibited by any two entities with sufficient regularities on any given scale" (2020, 3). She argues that this requires ontological commitment to *both* objects and real patterns.

Robust (i.e., projectible) patterns of relations among entities are indeed appropriately understood as real patterns. A good example is the case of the chemical bond. Seifert argues that chemical bonds are "real patterns of interactions between subatomic particles" (2023, 269). There is a way of computing the structure of a molecule that corresponds to the "bitmap," namely the description of the quantum state that is produced by solving the Schrödinger equation from first principles (Seifert 2023, 275), but this is not what chemists tend to do—there are more efficient ways of describing a molecule, by focusing only on the electrons that occupy the outer shell orbital of each atom, because those are the ones that determine how atoms bond (279).[7] The other electrons can be safely disregarded, not because they do not enter into any interactions, but because they do not contribute to bond formation—thus they are not part of that pattern. Thus, chemical bonds are a real component of scientific ontology even though they are not material objects (283). Similarly, in the biological case, predator-prey relationships are patterns of interactions among organisms within ecosystems, and phylogenetic trees are diachronic patterns of ancestry and descent among species. Both are no less real components of scientific ontology.

But the objects that enter into these relations can also themselves be understood as real patterns. Consider an individual tiger. The tiger is a prime example of a real pattern. It is a real, objective physical object (Wallace 2012, 48), even though it is not part of basic physical ontology—instead, it should be understood as "a pattern or structure in the physical state" (Wallace 2003, 92). Although in principle it might be possible to make predictions about tiger hunting patterns "in terms of atoms and electrons" (Wallace 2003, 92), in practice this would be impossible, due to the absurd amount of computational power that would be required. In this case, the atom-for-atom account corresponds to the "bitmap" description of the tiger, and the object "tiger" is a real pattern because describing it at this scale is far more efficient, in terms of predicting and explaining its behavior, than providing the atom-for-atom bitmap description. In fact, tiger hunting patterns would *not* be reliably predicted or explained in terms of atoms and electrons, even in principle, because the pattern of the tiger's behavior—and the tiger itself—does not exist at that scale.[8] In order to describe the macroscopic motion of the tiger, for example, not only would taking into account all the Brownian motions of all the individual atoms and

7. Dennett (1991) describes efficiency in terms of information compressibility. Ross (2000) and Ladyman and Ross (2007) further precisify this notion in information-theoretic terms. Seifert (2023) understands it in terms of the degrees of freedom of the relevant descriptions.

8. "At the quantum scale there are no cats" (Ladyman and Ross 2007, 199).

molecules within each cell not contribute anything relevant to the description; it would be actively unhelpful, as the pattern would be lost among the noise.

There is more than one pattern to be found here, however. Although the organism itself (i.e., the physical object) is a real pattern, it is not the case, pace Wallace, that "any pattern which behaves as a tiger" is a tiger (Wallace 2003, 93). On the one hand, the tiger is a living, functioning physical object (Hermida and Ladyman 2022), made of various physical, chemical, and biological components at various scales, and it inherits some of its structure from the existence, properties, and behavior of these components. But in addition, something is a tiger only if it is appropriately related to certain other organisms. In other words, its location on the phylogenetic tree is partly what makes something a tiger. As Reydon notes (2019), phylogenetic classification does not primarily serve the function of encoding information about the properties of organisms (though it does encode *probabilistic* information concerning, for example, the higher likelihood that the organism will share more properties with other organisms to which it is more closely related). What it primarily does is *locate* the organism on the phylogenetic tree—that is, on the structure of relations among all living organisms that ultimately share a common origin.

The phylogenetic tree can also be described as a real pattern. The "bitmap" description of this structure would be the information about all the tokogenetic (i.e., genealogical) relationships among all the organisms that ever lived, including every single reproductive event.[9] The phylogenetic tree, in contrast, describes only the phylogenetic relationships among species (i.e., lineages that have become separated through speciation). Just as the pattern "tiger" could theoretically (though not in practice) be recovered from the information of the atoms and electrons that compose it, but the reverse is not the case, the phylogenetic relationships could also theoretically be recovered from the information about all the tokogenetic relationships arising from individual reproductive events, if it were available—but the reverse is not the case. However, all this information would massively decrease the efficiency of the description. Once there are discontinuities—once the phylogenetic tree begins to branch and split into separate lineages—it immediately becomes more efficient to track the relations among the discontinuous parts of the tree, rather than attempt to track the complete genealogical relationships among all organisms. Note that this is not merely a practical or computational matter—it is a feature of the world that there are these discontinuities. The existence of the discontinuities—that is, the establishment of separately evolving lineages

9. Think of an interminable biblical sequence of "A begat B who begat C who begat D," but far less linear.

against the background of the universal relatedness of all organisms, is the reason that species and higher taxa are real (Wiley and Lieberman 2011, 4). They are real patterns in the structure of evolutionary history.

As would be expected, although phylogenetics is the main classification system in biology, other classifications also capture real patterns, and are also valid classifications insofar as they do so. To stay with the tiger example, classifying this organism as a "predator" allows ecologists to model the predator-prey relationships between the population of tigers and the populations of its prey species, a real pattern which again is more efficiently captured in this way than by painstakingly listing every single interaction between individual organisms in the ecosystem. Equally good considerations can, no doubt, be brought to bear on the case for the existence of the real patterns that can be captured by classifications that identify a particular organism as, for example, "male," "juvenile," "multicellular," "placental," "homeotherm," or "terrestrial," in terms of explaining its behavior, developmental processes, life-history constraints, reproductive strategies, and so on. The multiplicity of real patterns in the living world thus justifies taxonomic pluralism.

9.6 Conclusion

Biology is the poster science for taxonomic pluralism. There are different species concepts and various classifications that cut across phylogenetic groups. However, all of these exist for good reason—not because they cater to the different interests of researchers, but because there are multiple real patterns in the world. What each classification attempts to do, and what each valid classification does, when all goes well, is track a real pattern in the structure of the world. This is so even for classifications that do not classify things into natural kinds, or into kinds at all, as is arguably the case for phylogenetic systematics.

The goal of science is to unveil the real patterns that exist in the world. The "ontological furniture" revealed by science includes not only things (i.e., physical objects), but also processes, relations, and interactions. All of these encode some projectible regularities, and all can be understood as real patterns. Since the world is complex and highly diverse in what it contains, but also nomologically stable and teeming with patterns at various scales, we should expect there to be multiple valid scientific classifications. This would be so even for a perfectly objective science entirely devoid of human interests.

References

Allmon, W. D. 2016. "Studying Species in the Fossil Record: A Review and Recommendations for a More Unified Approach." In *Species and Speciation in the Fossil*

Record, edited by W. D. Allmon and M. M. Yacobucci, 59–120. University of Chicago Press.

Barraclough, T. G. 2019. *The Evolutionary Biology of Species.* Oxford University Press.

Bird, A., and E. Tobin. 2024. *Natural Kinds.* Edited by E. N. Zalta and U. Nodelman. The Stanford Encyclopedia of Philosophy (Spring 2024 Edition). https://plato.stanford.edu/archives/spr2024/entries/natural-kinds/.

Creţu, A. 2020. *Natural Kinds as Real Patterns: Or How to Solve the Commitment Problem for Perspectival Realism.* [Preprint]. https://philsci-archive.pitt.edu/id/eprint/16828.

Darwin, C. 1859. *On the Origin of Species.* John Murray. https://darwin-online.org.uk/content/frameset?itemID=F373&viewtype=text&pageseq=1.

Dennett, D. C. 1991. "Real Patterns." *Journal of Philosophy* 88 (1): 27–51.

Derraik, José G. B., J. W. Early, G. P. Closs, and K.J.M. Dickinson. 2010. "Morphospecies and Taxonomic Species Comparison for Hymenoptera." *Journal of Insect Science* 10 (1): 108.

Dupré, J. 1993. *The Disorder of Things: Metaphysical Foundations of the Disunity of Science.* Harvard University Press.

Dupré, J. 1999. "Are Whales Fish?" In *Folkbiology,* edited by D. L. Medin and S. Atran, 461–476. MIT Press.

Ereshefsky, M. 2022. *Species.* Edited by E. N. Zalta. The Stanford Encyclopedia of Philosophy (Summer 2022 Edition). https://plato.stanford.edu/archives/sum2022/entries/species/.

Ereshefsky, M., and T.A.C. Reydon. 2015. "Scientific Kinds." *Philosophical Studies* 172:969–986.

Franklin-Hall, L. R. 2015. "Natural Kinds as Categorical Bottlenecks." *Philosophical Studies* 172:925–948.

Frost, D. R., and A. G. Kluge. 1994. "A Consideration of Epistemology in Systematic Biology, with Special Reference to Species." *Cladistics* 10 (3): 259–294.

Havstad, J. C. 2021. "Complexity Begets Crosscutting, Dooms Hierarchy (Another Paper on Natural Kinds)." *Synthese* 198 (8): 7665–7696.

Hennig, W. 1966. *Phylogenetic Systematics.* University of Illinois Press.

Hermida, M., and J. Ladyman. 2022. *Living Objects.* https://philsci-archive.pitt.edu/id/eprint/21430.

Hull, D. L. 1967. "Certainty and Circularity in Evolutionary Taxonomy." *Evolution* 21 (1): 174–189.

Hull, D. L. 1978. "A Matter of Individuality." *Philosophy of Science* 45 (3): 335–360.

Khalidi, M. A. 1998. "Natural Kinds and Crosscutting Categories." *Journal of Philosophy* 95 (1): 33–50.

Khalidi, M. A. 2013. *Natural Categories and Human Kinds: Classification in the Natural and Social Sciences.* Cambridge University Press.

Khalidi, M. A. 2024. "Ontological Pluralism and Social Values." *Studies in History and Philosophy of Science* 104:61–67.

Khalidi, Muhammad Ali. 1993. "Carving Nature at the Joints." *Philosophy of Science* 60 (1): 100–113.

Ladyman, J., and D. Ross. 2007. *Every Thing Must Go*. Oxford University Press.

Ludwig, D., and S. Ruphy. 2021. *Scientific Pluralism*. Edited by E. N. Zalta. The Stanford Encyclopedia of Philosophy (Winter 2021 Edition). https://plato.stanford.edu/archives/win2021/entries/scientific-pluralism/.

Mallet, J. 1995. "A Species Definition for the Modern Synthesis." *Trends in Ecology & Evolution* 10 (7): 294–299.

Mayr, E. 1942. *Systematics and the Origin of Species*. Columbia University Press.

Mayr, E. 1995. "Species, Classification, and Evolution." In *Biodiversity and Evolution*, edited by R. Arai, M. Kato, and Y. Doi, 3–12. National Science Museum Foundation.

McGhee, G. R. 2011. *Convergent Evolution: Limited Forms Most Beautiful*. MIT Press.

Okasha, S. 2002. "Darwinian Metaphysics: Species and the Question of Essentialism." *Synthese* 131:191–213.

Plutynski, A. 2018. *Explaining Cancer: Finding Order in Disorder*. Oxford University Press.

Queiroz, K. de. 2007. "Species Concepts and Species Delimitation." *Systematic Biology* 56 (6): 879–886.

Queiroz, K. de, and M. J. Donoghue. 1988. "Phylogenetic Systematics and the Species Problem." *Cladistics* 4 (4): 317–338.

Reydon, T. A. C. 2019. "Taxa Hold Little Information About Organisms: Some Inferential Problems in Biological Systematics." *History and Philosophy of the Life Sciences* 41 (4): 40.

Richards, R. A. 2016. *Biological Classification: A Philosophical Introduction*. Cambridge University Press.

Ross, D. 2000. "Rainforest Realism: A Dennettian Theory of Existence." In *Dennett's Philosophy: A Comprehensive Assessment*, edited by D. Ross, A. Brook, and D. Thompson, 147–168. MIT Press.

Ruphy, S. 2016. *Scientific Pluralism Reconsidered: A New Approach to the (Dis)Unity of Science*. University of Pittsburgh Press.

Seifert, V. A. 2023. "The Chemical Bond Is a Real Pattern." *Philosophy of Science* 90 (2): 269–287.

Simpson, G. G. 1961. *Principles of Animal Taxonomy*. Columbia University Press.

Sokal, R. R., and P. H. A. Sneath. 1962. *Principles of Numerical Taxonomy*. W. H. Freeman.

Stankowski, S., and M. Ravinet. 2021. "Quantifying the Use of Species Concepts." *Current Biology* 31 (9): R428–R429.

Tobin, E. 2010. "Crosscutting Natural Kinds and the Hierarchy Thesis." In *The Semantics and Metaphysics of Natural Kinds,* edited by H. Beebee and N. Sabbarton-Leary, 187–199. Routledge.

Wallace, D. 2003. "Everett and Structure." *Studies in History and Philosophy of Science Part B: Studies in History and Philosophy of Modern Physics* 34 (1): 87–105.

Wallace, D. 2012. *The Emergent Multiverse: Quantum Theory According to the Everett Interpretation.* Oxford University Press.

Wiley, E. O., and B. S. Lieberman. 2011. *Phylogenetics: Theory and Practice of Phylogenetic Systematics.* 2nd. Wiley Blackwell.

Zachos, F. E. 2016. *Species Concepts in Biology.* Springer.

Zink, R. M., and J. Klicka. 2018. "Species Concepts and Speciation Analysis." In *Ornithology: Foundation, Analysis, and Application,* edited by M. L. Morrison, A. D. Rodewald, G. Voelker, M. R. Colón, and J. F. Prather, 39–60. JHU Press.

10 Deciphering the Noise: Real Patterns in Welfare from Incentivized Choice

Aleksandr Alekseev, Glenn W. Harrison, Morten Lau, and Don Ross

10.1 Introduction

A persistent challenge in extending Dennett's (1991) account of real patterns (RPs) across the empire of the sciences is that the motivations of the account were rooted in one province of that heterogeneous domain, cognitive psychology. The driving goal of Dennett's argument was to explain how one science could be about both the propositional attitudes of folk psychology *and* about information processing in nervous systems, without trying to either explain away or reduce the first kind of pattern to the second kind. Some aspects of this problem are *sui generis*, and to that extent "Real Patterns" is an exercise in the philosophy of psychology; other aspects of it generalize, and where those are concerned, Dennett's account is a major contribution to the philosophy of science. Complicating this problem further is that the demarcation between special and general aspects will vary across the disciplines to which the idea of an RP is extended.

One of us has dealt elsewhere with elements of the RP ontology that apply specifically to economics (Ross 1994, 2009). However, these studies address only so-called "positive," or "descriptive," aspects of economics, where economics arguably differs from other mathematical behavioral sciences only with respect to the highest-level principle by which economists identify patterns of special interest to them, namely that such patterns be generated from interactions between at least one agent and some contingent incentive. "Normative" economics—which to most economists is synonymous with the economics of welfare—raises special issues, which are related to aspects of Dennett's original application of RP theory in ways that, to our knowledge, have not previously been treated outside of the home context in cognitive science.

A primary goal of "Real Patterns" was to argue for the reality—on grounds of indispensability to *scientific*, and not just quotidian, explanation—of the patterns tracked from what Dennett's earlier work (1989) had called "the intentional stance" (IS). The basic idea of the IS is that beliefs, preferences, and other propositional attitudes are attributions, to patterns of behavior, of indexing content (e.g., "believes that bees are important to human food supplies," "hopes that bee population decline can be arrested") that consolidates the patterns in question under publicly accessible *rationales*. That is, ascriptions of propositional attitudes should not be understood as hypotheses about hidden "mental states" of ascribees, but rather as summaries of an ascribee's probable histories of information exposure, problem solving, social sensitivity, and planning. IS ascriptions thus aim to capture and interweave far more *kinds* of information than concern a neuropsychologist studying a person's (or other animal's) neural processing of a cognitive function.

From an economist's point of view, IS attributions are thus coordinating devices that allow agents to identify their actual and potential strategic interactions, which form the basis of a society (as opposed to a mere aggregation of individuals). The core of becoming socialized is for a person to learn to apply the IS to themselves, and to integrate into their entire suite of explicit thought and behavior a normative obligation to be comprehensible from the IS. A crucial part of the scientific evidence for this model is that children learn to apply the IS to others *before* they can successfully apply it to themselves (McGeer 2002, 2001). Children do not introspectively discover beliefs and preferences in themselves, and then hypothesize that others privately experience similar states. Thus, people can be wrong about their intentional self-attributions, and correctable by others, when there are observable and relevant patterns in their behavior (including their verbal behavior) that they have not noticed. This answerability to standards other than the agent's subjective impressions, along with the fact that ascriptions from the IS are sensitive to ecological problem contexts and not just internal information-processing states of ascribees, are what make the patterns tracked from the IS irreducibly "real," according to Dennett (1991).

The aspect of the IS not explicitly taken up in previous extensions of the RP ontology beyond cognitive science is its *normative* element. Beliefs and preferences, according to the IS, are ascribed, and self-ascribed, on the basis of what agents "ought to" believe and prefer if they are responsible to standards of comprehensibility. A person can report a belief that no one else shares, or such a belief can be ascribed to her, if accompanied by acknowledgment of some idiosyncratic circumstance ("she is a genius"; "she is thought to commune with the spirit world"; "she was the only witness to the crime"). But attribution of beliefs or preferences that are contradictory on their surfaces, and not intended figuratively ("she believes that Zeke is a married bachelor"; "she prefers hockey to baseball and baseball to hockey") run afoul of the point of

IS attribution. In cases where an agent systematically proclaims obvious contradictions, then we doubt that she satisfies the patterns of agency (at least in the domain of her incomprehensible utterances), suspend the IS with respect to these behaviors, and search for an explanation in terms of features of her nervous-system profile (for example, "she recently suffered a severe stroke").

This restriction based on norms of consistency does not apply to RPs in sciences that are not about agents, including sciences that apply to aspects of people not directly related to their agency. It is relevant to RPs in descriptive economics, insofar as there would be no economic patterns of any kind in a world where no agent was generally relatively consistent in their responses to incentives. But, at least in principle, identification of descriptive economic RPs does not require mappings between coherent demand and supply curves and specific agents who are taken to be trying to optimise personal utility, so the strong normative assumptions of the IS can be massively relaxed (Ross 2012; Hildenbrand 1994; Becker 1962).

Where welfare economics is concerned, however, the situation more closely resembles that of individual and social psychology understood according to the IS. Welfare economists have historically treated preferences, along with beliefs about probabilities, as exogenous phenomena, or, better to say, as RPs that are inferred from agents' choice behavior. Though this assumption can arguably be relaxed in special, motivated conditions such as addiction, it is not negotiable in general because the very point of welfare economics is to identify policies that satisfy preferences as efficiently as binding circumstantial constraints allow. Preferences—along with beliefs—interact with one another both within and between agents, so every exercise in welfare economics is a study of trade-offs, specified on the basis of identifying relatively consistent *sets* of preferences and beliefs associated with specific agents affected by policies under assessment. Our focus here is mainly on preferences, without ignoring the fact that beliefs systematically interact with risk preferences except when laboratory task designs deliberately separate them.

Economists represent sets of an agent's preferences by utility functions that map relative costs and risks associated with options onto optimizing choices. There are various models of utility that economists apply to different problems, but all such models specify consistency constraints to which preferences must conform if the relevant functional form specified by the utility model in question is to be identifiable.

Utility and choice patterns are not specified independently; rather, utility models structure observable choice patterns. Thus, as argued by Ross (2005, 2014) and Harrison and Ross (2023), economists' "revealed preference theory" (RPT), which "rationalizes" actual choices in such a way as to identify both optimizing structures demonstrated by agents' choices and actual sets of optima in potentially available choice sets, is in fact a mathematical operationalization of the IS. Harrison and Ross (2023) consequently call their methodological program for behavioral welfare economics, which incorporates a specific approach to RPT that accommodates empirically observed patterns of choice under risk, "the quantitative intentional stance" (QIS). Since RPT has its original origins in the 1950s, economists were effectively Dennettians before Dennett came along to provide a philosophical account of their activity that was superior to any they had provided themselves.

There are many specific RPT models, sculpted for different purposes of application (Chambers and Echenique 2016). Prior to development of the general theory of revealed preference (GARP) by Afriat (1967), RPT amounted to a formal philosophy of consumer demand rather than an operational tool, because theorems were existence proofs applied to infinite choice sets, to which response cannot be observed. When we refer to RPT in this chapter, we should be read as referring to approaches such as GARP that identify utility functions, when they exist, in finite, empirically observable, sets of choices.

We have now presented enough background to be able to state the objective of the chapter. Most work on the philosophy of RPs has sought to make the idea more rigorous by grounding it in algorithmic computational or information theory. Such work is well represented in this book. Approaches of this kind, which originate in work by Ladyman and Ross (2007, chap. 4) aim to capture *epistemic* norms by which RPs are distinguished from merely apparent (not projectible) or redundant (reducible) patterns. But they do not advance the project of providing quantitative structure, facilitating measurement in inference tests and experimental settings, of Dennett's original model of RPs, the patterns tracked from the IS.

RPT models sometimes identify utility functions conditional on the assumption that every discrete choice made by an agent is either the optimal one for that agent, or is an error. This is an extremely strong restriction, which can be motivated in cases where strongly institutionally constrained agents, for example corporations employing professional analysts, make choices implicating simple utility functions such as attempted maximization of expected monetary profits. Economists, however, study limitless ranges of choice contexts where natural agents such as people and other animals have limited information, limited attention, and ambivalent tastes. All economists who conduct experiments with human subjects know that observed choice patterns involve stochastic influences, so that versions of RPT they apply must assume agents whose choices respect only stochastic rather than strict dominance.[1] Considerable research in economics is devoted to explaining this behavior.[2] Were this not the case, there would be little motivation for regarding the theoretical objects of study as patterns rather than states governed by field equations.

Well-studied *descriptive* models of stochastic choice provide econometric tools to estimate structural models in a broad range of applications, including applications that are positive and descriptive rather than normative. The primary engineering spec for such tools is that they recover the structural parameters of the deterministic part of a model, reflecting risk or time preferences. Little attention has been given, however, to the systematic economic interpretation of the parameter estimates of the stochastic parts of such models, which determines the magnitude of choice randomness. The estimated structural parameters of such models also have a stochastic component, such as standard errors for classical estimation methods or standard deviation for Bayesian estimation methods. As will become clear, our reference to the "stochastic part of the model" is to additional parameters used to characterize random choice behavior. The interpretation of these parameters allows one to characterize the economic value of choice randomness, which is essential for applications to welfare assessment.

In light of this characterization of the state of the literature, this chapter is more than a quantitative extension of the IS. It is, in addition, a practical contribution to economic methodology.

One way of understanding what an economist making welfare assessments is doing is that she is using her knowledge of some inferred RPs to generate priors to inform inferences about welfare gains and losses from newly observed choices. One way in which these priors can be generated is by observing the same agent make decisions in some other setting than the choices of normative interest. In Harrison and Jia Min Ng (2016) and Harrison and J. M. Ng (2018) and Gao et al. (2023), the domain of normative interest was decision-making over insurance products, and the observed patterns on which priors were formed were risk preferences estimated from prior experiments with the same subjects. Similarly, in Harrison and Ross (2018), experimental elicitation of risk preferences was used to build priors for normatively evaluating choices over investment portfolios in a subsequent experimental stage.

1. Nogee and Mosteller (1951) provide the earliest evidence of stochastic choice, followed by Tversky (1969), Starmer and Sugden (1989), Camerer (1989), and Ballinger and Wilcox (1997).

2. Wilcox (2008) provides an excellent overview of many popular stochastic models of choice under risk. Recent examples include Swait and Marley (2013), Wallin et al. (2018), Matêjka and McKay (2015), and Agranov and Ortoleva (2017).

While we think that this method for estimating welfare effects can often be taken up with good epistemic confidence, it is always open to skepticism about the transferability of inferred patterns across specific experiments. Here, we therefore develop a complementary approach that is immune to such skepticism, and which allows us to exhibit some strong technical properties of the QIS. We seek to form priors from some choices in an experimental task that allow us to infer welfare consequences of other observed choices in that *same task.*

Consider a generic structural model of discrete choice that uses a standard multinomial logit model of stochastic choice, also known in the literature as the strong utility model or the Fechnerian model, which assigns each discrete alternative a choice likelihood $\mathbb{P}$ according to

$$\mathbb{P}(a \mid \beta, \mu) = \frac{\exp(U(a \mid \beta)/\mu)}{\sum_{a' \in A} \exp(U(a' \mid \beta)/\mu)}. \tag{10.1}$$

In this expression, a and a' are alternatives, such as lotteries or dated outcomes, from a set of all alternatives A. The deterministic part of this structural model is parameterized by a vector of behavioral parameters β, which could represent, for instance, an agent's risk or time preferences. For example, in the case of risk preferences, β could be a risk aversion parameter and U could be the expected utility of a risky alternative; in the case of time preferences, such as the quasi-hyperbolic discounting model, β would comprise the exponential and hyperbolic discounting parameters and U would be the discounted utility of an income stream. The behavioral parameters determine the aggregate utility function U used to model each alternative.

The stochastic part of the model is parameterized by μ, often called the *noise* parameter.[3] The noise parameter determines how sensitive choice likelihoods are to the maximization of utility U according to a given structural model. As noise tends to zero, an agent will almost surely choose the alternative with the highest utility. When noise goes to infinity, the agent will assign equal likelihoods to choosing each alternative regardless of their utilities. Higher values of μ thus imply a higher magnitude of choice randomness in this popular specification.

Three issues arise with the interpretation of the estimates of the noise parameter. First, while the effect of μ on choice likelihoods is clear, the noise element

3. In the game theory literature on "quantal response equilibrium" due to McKelvey and Palfrey (1995), which applies stochastic choice to strategic settings, it is common to use an alternative parameterization $\lambda \equiv 1/\mu$.

has no economic interpretation in itself.[4] A monetary value assigned to a noise estimate that reflects forgone welfare, on the other hand, would provide information about economic consequences of choice randomness. Second, since the noise parameter is unbounded from above, it is difficult to judge whether the randomness of an agent's choices is high or low. A bounded value, such as one defined on the unit interval would, solve this problem.[5] Third, the raw estimates of μ are not well suited for interpersonal welfare comparisons, of the kind required for most interesting policy evaluations, since behavioral parameters β also change across people. Having choice randomness expressed in common units, such as the monetary value of the certainty equivalent of foregone welfare, and taking into account the interpersonal differences in β could help to overcome this limitation.[6]

We address these issues by converting an estimate of μ into two intuitive measures.[7] The first measure, absolute welfare cost (AWC), puts a dollar value on choice randomness: It shows how much money, in certainty equivalent terms, an agent would be allowed to "waste" consistent with rationalization of her choices by an underlying structural model.[8] The second measure, relative welfare cost (RWC), scales the absolute welfare cost by certainty equivalent of the monetary value at stake in a choice context. The relative welfare cost is thus defined on the unit interval. It shows what proportion of the total value at stake an agent could be interpreted as potentially "wasting" consistent with rationalization of her choice by the model. That is, we could capture the normative leverage gained, in a specific case, by applying the QIS perspective to analysis.

4. In the existing literature (Gaudecker et al. 2011; Bland 2020), an estimate of noise is sometimes interpreted as the likelihood of choosing the best alternative (between the two available) for a given difference in utilities (or certainty equivalents) between them. While this number is informative of the economic consequences of choice randomness, it does not provide a measure of the welfare costs associated with stochastic choice.

5. The parameter of the tremble model of stochastic choice (Harless and Camerer 1994) has this property and thus allows one to evaluate the relative magnitude of choice randomness. However, an estimate of the tremble parameter still requires an economic interpretation. See Carbone and Hey (2000) for a comparison between the tremble model and the Fechnerian model.

6. Aspects of these three issues arise not only in the standard multinomial logit model but also in its modifications, such as the contextual utility model of Wilcox (2011) or specifications that substitute certainty equivalents of alternatives for their expected utilities, such as Gaudecker et al. (2011).

7. While the discussion below focuses on the multinomial logit model and its modifications, a similar logic can be applied to other models of stochastic choices, such as the trembles model (Harless and Camerer 1994) or the random preferences model (Loomes and Sugden 1995; Gul and Pesendorfer 2006).

8. While our discussion focuses on individual decision-making, our method can also be used to study stochastic choice in group decision-making (Bone et al. 1999).

Our approach thus rests on a careful basis for distinguishing "noise" from "waste." Noise, a property of a real choice pattern, creates cost *against which* choices can be assessed as being more (welfare opportunity capturing) or less (wasteful) efficient. Some literature in economic methodology has anticipated the spirit of the IS in understanding the RPs that the welfare analyst aims to isolate. Infante, Lecouteux, and Sugden (2016, 21) say that the kind of estimation we seek is

> not an inference about the hypothetical choices of the client's inner rational agent, but rather a way of *regularizing* the available data about the client's preferences so that it is compatible with the particular model of decision-making that the professional wants to use. Regularization in this sense is almost always needed when a theoretical model comes into contact with real data.

As often in welfare economics, here the subject being evaluated is the "client," and the economist performing the analysis, intended to provide the basis for advice, is the "professional."

We apply our method to data from an artefactual field experiment conducted in Denmark. The subjects came from a sample of the general Danish population and were asked to make a series of choices between two risky alternatives. Each subject answered a detailed demographic survey, which we use to characterize the effects of demographic characteristics on the observed heterogeneity in the AWC and RWC. We find that the average AWC are around 67 Danish kroner ($10)[9] and thus negligible for the subjects' natural economic environment. However, the RWC are quite significant, at 0.87 on average. There is also considerable variation among the subjects in terms of their AWC and RWC. Descriptive regression analysis shows that certain demographic characteristics, such as age, education level, employment status, and sex are associated with higher welfare costs from noisy choice patterns.

The chapter is organized as follows. Section 10.2 describes the method of converting an estimate of noise into welfare costs measured in monetary terms and provides an explicit algorithm for computation in a binary choice case. Section 10.3 applies the method to data from the Danish field experiment and studies properties of the estimated welfare costs. Section 10.4 discusses connections with previous literature in economics, particularly aimed at helping non-economists see the value added by the QIS to economists' ambitions. Section 10.5 offers concluding reflections.

9. Throughout the text, we use an exchange rate of 1 Danish krone = $0.15 that was prevalent at the time of the experiment.

10.2 Method

We first look at a general case when the set of alternatives is continuous. This case allows us to demonstrate the logic behind our method of extracting the welfare cost information from a noise estimate. Then we turn to a more common discrete case with two alternatives and explicitly describe the algorithm to implement our method.

10.2.1 General Case

Consider an agent choosing from a set of alternatives indexed by real numbers on a compact interval $A = [a_l, a_h]$. Each alternative generates a lottery[10]

$$l(a) = \{x_1(a), \ldots, x_k(a);\, q_1(a), \ldots, q_k(a)\},$$

$$a \in A,\ x_i \in \mathbb{R},\ q_i \in [0, 1],\ \forall i = 1, \ldots, k,\ \sum_{i=1}^{k} q_i = 1,$$

where x_i are monetary outcomes and q_i are respective probabilities of obtaining those outcomes.

This setting could represent allocating resources between two state-contingent accounts, as in Choi et al. (2007). Each allocation in this example is an alternative with $k = 2$ outcomes, $x_1(a)$ and $x_2(a)$, and equal probabilities of each outcome. The minimum ($a_l = 0$) and maximum ($a_h > 0$) amounts an agent can allocate to account 1 will define the interval of alternatives A. Then $x_1(a) = a$ and $x_2(a) = b(a_h - a)$, where $-b < 0$ is the slope of the budget line and $q_1(a) = q_2(a)$, $\forall a \in A$.[11]

The risk elicitation task of Gneezy and Potters (1997) is another example of such a setting. In this example, the minimum ($a_l = 0$) and maximum ($a_h > 0$) amounts a subject can allocate to a risky asset define the set of alternatives A, where a_h is the initial endowment. A subject's choice of how much of the endowment to allocate to a risky asset, a, generates lotteries with two outcomes given by $x_1(a) = a^h - a$ (the asset yields no return) and $x_2(a) = a^h + a(k-1)$ (the asset yields a positive return $k-1$). The probabilities of outcomes are given exogenously and do not depend on a.

Each alternative a has an expected utility $EU(a) \equiv EU(l(a))$ defined by an assumed structural model of choice under risk. Although we focus on the expected utility (EUT) model here, our approach can be applied to other models of choice under risk that capture frequently observed patterns, and online

10. The lottery itself does not need to be discrete. An alternative can generate a continuous probability density.

11. In an actual experiment, the set of alternatives is, of course, discrete. This choice set, however, comes close to being continuous.

appendix C illustrates the extension to "rank dependent utility" (RDU) models of risk preferences.[12]

Monetary outcomes are transformed using $u:\mathbb{R}\mapsto\mathbb{R}$, the von Neumann-Morgenstern utility function. Each value of $EU(a)$ can be translated into a certainty equivalent $m(a)$, defined by $u(m(a))=EU(a)$. For non-economists, the certainty equivalent is just the amount of money for certain that is equal in utility to the utility provided to the agent from owning the risky asset a. The ordering of alternatives is preserved for the certainty equivalent transformation: $EU(a)\geqslant EU(b)\Leftrightarrow m(a)\geqslant m(b),\ \forall a,b\in A$.

Assume that EU is concave and reaches its unique maximum (minimum) at a^* (a_*), as does the certainty equivalent function. Define the maximum certainty equivalent as $m^*\equiv m(a^*)$, and the minimum certainty equivalent as $m_*\equiv m(a_*)$. If the agent always chooses the best alternative a^*, we call this behavior *perfectly rationalizable* (by an assumed model of choice under risk). At the other extreme, if the likelihood of choosing a^* is the same as for any other alternative, we call such a behavior *non-rationalizable*. We are concerned with the behavior in between, which is neither perfectly rationalizable nor non-rationalizable, a behavior that we call *imperfectly rationalizable*. Crucially with respect to the point of our project, this expression should be understood as a measure of the inability of a given model to regularize data, rather than a statement about an agent making decision errors. That is, our interest here is in identifying patterns, not in hypothesizing mechanisms that might generate them.

The degree of this imperfect rationalizability can be characterized by a number ε, $0\leqslant\varepsilon\leqslant\Delta m$, with $\Delta m\equiv m^*-m_*$. Choices that lead to certainty equivalents within ε distance from the maximum certainty equivalent can be viewed, from the perspective of a model, as imperfectly rationalizable.[13] These choices form an optimal region A^* defined by

$$A^*(\varepsilon)=\left\{a\in A\mid m(a)\geqslant m^*-\varepsilon\right\}. \tag{10.2}$$

The degree of imperfect rationalizability ε shows how much monetary-equivalent welfare an agent would be allowed to waste to make her choices rationalizable by the model, and effectively includes these choices in the optimal region. In other words, ε represents the welfare costs measured in monetary units. Our goal is to link these costs to noise.

12. Online appendices are available in Alekseev et al. (2024) at https://cear.gsu.edu/category/working-papers/wp-2024/.

13. The idea of estimating models that tolerate imperfect rationalizability of choices is not new. For example, Harrison (1994) introduces a similar quantity based on an agent's subjective cost of choosing one alternative versus the other to explain many violations of expected utility theory.

The allowed degree of imperfect rationalizability covaries with the width of the optimal region. If ε is set to 0, the optimal region will consist only of the best alternative a^*. If $\varepsilon = \Delta m$ the optimal region will coincide with the whole set of alternatives A. Figure 10.1 illustrates how the optimal region varies with the degree of imperfect rationalizability. Geometrically, the optimal region is the line segment $\left[a_l^*, a_h^*\right]$.

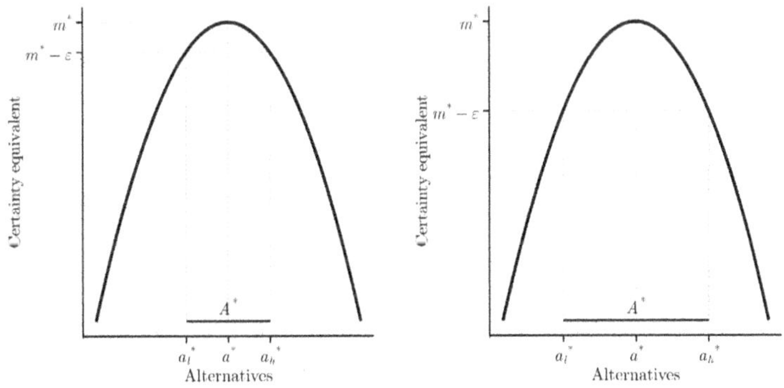

(a) Low imperfect rationalizability (b) High imperfect rationalizability

Figure 10.1: The optimal region and degree of imperfect rationalizability.

The optimal region and the degree of imperfect rationalizability are the first two components that we need to interpret an estimate of noise. The third component comes from a stochastic model $p: A \mapsto \mathbb{R}_+$, which generates choice likelihoods over the set of alternatives. Some alternatives fall into the optimal region, by definition. By integrating the density $p(a)$ over this region we get the proportion of choices that are counted, from the perspective of a model, as imperfectly rationalizable. We call this measure a *degree of rationalizability* (DoR):

$$\rho(\mu, \varepsilon) = \int_{A^*(\varepsilon)} p(a)\mathrm{d}a. \tag{10.3}$$

The DoR has several intuitive properties, two of which turn out to be crucial for our analysis, and can be represented graphically. Figure 10.2 shows that as the degree of imperfect rationalizability increases, the optimal region expands and the DoR, represented by the gray shaded area, increases. Figure 10.3 shows that as the noise goes up, the density flattens out and the probability mass shifts from the optimal region to the outside area, reducing the DoR.

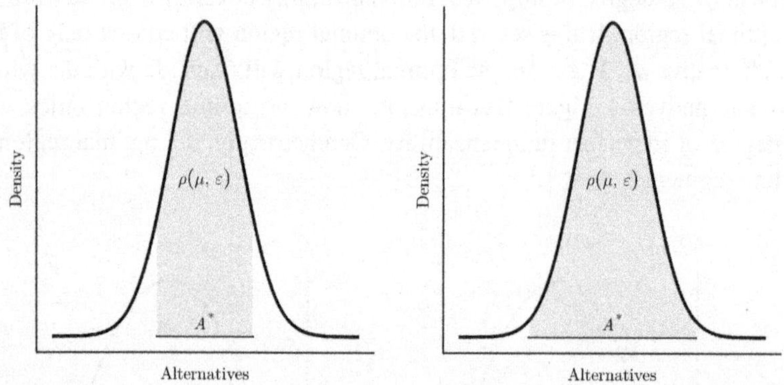

(a) Low imperfect rationalizability (b) High imperfect rationalizability

Figure 10.2: Degree of rationalizability and imperfect rationalizability.

The DoR for certain values of noise and imperfect rationalizability has attractive interpretations. The quantity $\rho(\infty, \varepsilon)$ tells us what proportion of choices are counted as rationalizable for a given imperfect rationalizability ε when they are, in fact, non-rationalizable. It represents a Type II error in a test to detect rationalizability, and the quantity $1-\rho(\infty, \varepsilon)$ is the power of this test. This power will decrease as the allowed degree of imperfect rationalizability increases or as the set of alternatives shrinks. The value of DoR at $\rho(\hat{\mu}, 0)$ measures the proportion of rationalizable choices for an estimated level of noise $\hat{\mu}$ and no imperfect rationalizability. We refer to it as the *default degree of rationalizability* or DDoR.

We now have all the tools to decipher the noise. We do this by linking an estimate of μ, the value of which is hard to interpret directly, to the degree of imperfect rationalizability, a monetary measure that has an intuitive economic interpretation as the welfare cost, or monetary-equivalent welfare required to rationalize the agent's choices by a model. In order to link them, we need to reverse the steps we followed so far. Currently, we introduced a degree of imperfect rationalizability ε that defines an optimal region A^*. The optimal region combined with a stochastic model, parameterized by μ, yields a value of DoR. Now suppose that instead we start with a DoR measure and fix it at some target level α. Let an estimated value of the noise be $\hat{\mu}$. The question is how much imperfect rationalizability should be allowed for $100 \times \alpha\%$ of the choices to be rationalized for a given noise. In other words, we need to find ε that satisfies

$$\rho(\hat{\mu}, \varepsilon) = \alpha. \tag{10.4}$$

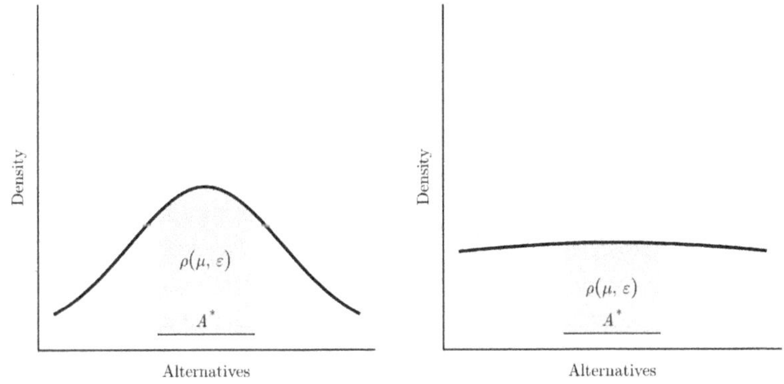

(a) Low noise (b) High noise

Figure 10.3: Degree of rationalizability and noise.

This equation establishes an implicit function, $\varepsilon(\hat{\mu};\alpha)$. For the purpose of our analysis, the following property of this function is important:

Proposition 1. *For a given α, the degree of imperfect rationalizability as a function of noise, $\varepsilon(\mu;\alpha)$, is monotonically increasing:*[14]

$$\frac{d\varepsilon}{d\mu} \geqslant 0.$$

Proof. See appendix 10.6.

This property implies that noise and imperfect rationalizability are in a direct and monotonic relation.[15] This property is important since more noise should imply higher welfare costs, which in our case are measured by imperfect rationalizability. If imperfect rationalizability and noise were not in a direct and monotonic relation, such an interpretation would be impossible. The relation between ε and μ comes from the fact that the DoR is decreasing in noise and increasing in imperfect rationalizability. From these properties it also follows that higher values of α imply higher values of ε. The more choices we wish to rationalize, for a given value of a noise, the more imperfect rationalizability we

14. We note that when alternatives are discrete rather than continuous, as discussed below, the DoR as a function of imperfect rationalizability will not be continuous and thus it will not be possible to match the target DoR α perfectly. We address this issue by using a discrete grid for ε and an interpolated version of $\rho(\hat{\mu},\varepsilon)$. The interpolated DoR function on a discrete grid is continuous and thus proposition 1 applies.

15. This property holds for a given agent, or rather for given risk preferences. This property will not hold perfectly across agents whose preferences are different.

should allow. The choice of the target α is left to the discretion of a researcher. In our empirical analysis we use the value of 0.95.

So far we have focused on a single choice context, but in practice we observe agents making choices over a series of rounds of a choice task. Suppose we observe an agent's choices c_j over n rounds indexed by $j = 1, \ldots, n$, and in each round the mapping of alternatives a into lotteries $l_j(a)$ is different. In the context of an allocation task, for example, the variation is introduced by changing the slope of a budget line b^j and a maximum amount a_h^j that can be allocated to one of the accounts: $A_j = [0, a_h^j]$, $x_1^j(a) = a$, $x_2^j = b^j(a_h^j - a)$. We assume that μ and preferences remain fixed for the duration of a choice task. We also keep the degree of allowed imperfect rationalizability ε the same across all rounds. We could repeat all the previous steps in deriving the DoR, although now it would differ by the round, $\rho_j(\mu, \varepsilon)$, and then aggregate these individual DoR from all the choices by taking their average.

With multiple choices, however, we can compute the DoR more directly by looking at the proportion of actual choices that fall into the optimal region:

$$\rho(\mu, \varepsilon) = \frac{1}{n} \sum_{j=1}^{n} \mathbb{I}\left[c_j \in A^*(\varepsilon)\right]. \tag{10.5}$$

Naturally, the average follows all the properties of the DoR for a single choice. In particular, it increases in ε and decreases in μ. We can then use the aggregate DoR in (10.5) to calculate the imperfect rationalizability needed to reach a target α in equation (10.4).

After calculating the value of ε that satisfies equation (10.4), $\varepsilon(\hat{\mu}, \alpha)$, it makes sense to adjust this value to take into account the fact that the degree of imperfect rationalizability should not exceed the difference between the maximum and minimum certainty equivalents for a given choice.[16] Since a common ε is applied to all the rounds of a choice task, for some rounds it can actually exceed Δm. Increasing imperfect rationalizability beyond this difference does not have any effect on the DoR and would imply that we allow an agent to waste more monetary-equivalent welfare than there actually is. This issue can be addressed by bounding ε by Δm, and then averaging across all the rounds:

$$\bar{\varepsilon}(\hat{\mu}, \alpha) = \frac{1}{n} \sum_{j=1}^{n} \min\left\{\varepsilon(\hat{\mu}, \alpha), \Delta m_j\right\}. \tag{10.6}$$

16. If payoffs differ substantially across lottery pairs, as in the Holt and Laury (2002) 20x and 50x treatments, one would want to use a contextual utility model (Wilcox 2011) or an expo-power utility function to accommodate such scale changes. We show that our results are robust to using these alternative specifications in online appendix C.

We call the resulting measure of imperfect rationalizability *absolute welfare costs* (generated by noise $\hat{\mu}$, with $100 \times \alpha\%$ of choices rationalized), or AWC. It represents the monetary-equivalent welfare that the agent would be allowed to give up for exactly $100 \times \alpha\%$ of her choices to be rationalized by the model, given noise $\hat{\mu}$. For any estimated value of noise and any desired proportion of choices we would like to rationalize we can, therefore, always find a precise dollar value equivalent of the welfare costs.

We can go further and translate the welfare costs into relative terms, to compare these costs with the actual stakes of a choice context. For example, an AWC of \$1 may not seem substantial, but if Δm_j are close to \$1 in each of the rounds, almost all of the possible welfare would have to be sacrificed to rationalize an agent's choices. We divide the degree of imperfect rationalizability by the difference between the maximum and minimum certainty equivalents for every round and average across all the choices (since the resulting quantity has to be a fraction, we bound the ratio by 1):

$$\tilde{\varepsilon}(\hat{\mu},\alpha) = \frac{1}{n}\sum_{j=1}^{n} \min\left\{\frac{\varepsilon(\hat{\mu},\alpha)}{\Delta m_j}, 1\right\}. \tag{10.7}$$

The resulting degree of imperfect rationalizability represents *relative welfare costs* (generated by noise $\hat{\mu}$, with $100 \times \alpha\%$ of choices rationalized), or RWC. Another benefit of this measure is that it allows one to appreciate the relative magnitude of noise, since RWC are bounded between 0 and 1, while a raw estimate of noise is unbounded from above. If rationalizing $100 \times \alpha\%$ of the choices requires on average almost all the difference between the maximum and the minimum certainty equivalents, so that the RWC are close to 1, that indicates that the choices are close to being non-rationalizable from the perspective of the model. On the other hand, if it requires only a small fraction of this difference, so that the RWC are near 0, then choices are close to being perfectly rationalizable from the perspective of the model.

10.2.2 Binary Choice

An important special case arises when an agent has only two alternatives to choose from. This is one of the most common experimental designs in risk elicitation tasks. In this case the set of alternatives in each round is $A = \{a_1, a_2\}$. Without loss of generality, assume that alternative a_2 always gives the highest expected utility, so that $EU_j(a_2) > EU_j(a_1)$, $j = 1, \ldots, n$—that is, $a_j^* = a_2$, using the notational convention $EU_j(a) \equiv EU\left(l_j(a)\right)$. The maximum and the minimum certainty equivalents in each round j are $m_j^* \equiv m_j(a_2)$ and $m_{j*} \equiv m_j(a_1)$, respectively. The optimal region can then take only two values:

$$A_j^*(\varepsilon) = \begin{cases} a_2, & \varepsilon < \Delta m_j, \\ A, & \varepsilon \geqslant \Delta m_j. \end{cases} \tag{10.8}$$

Suppose we observe a series of binary choices c_j made by a subject and estimate a structural model of risk preferences in which $\hat{\gamma}$ is a vector of estimated risk parameters and $\hat{\mu}$ is an estimate of noise. The $\hat{\gamma}$ vector in the EUT case is typically just the relative risk aversion. In the case of RDU $\hat{\gamma}$ includes the risk aversion parameter(s) and the probability weighting parameter(s). The computation of AWC and RWC (rationalizing $100 \times \alpha\%$ of the choices) from these data can be performed using the following algorithm:

1. For each round, compute the expected utilities of both alternatives, $EU_j(a_1; \hat{\gamma}), EU_j(a_2; \hat{\gamma}), j = 1, \ldots, n$.
2. Compute the certainty equivalents of both alternatives, $m_j(a_1), m_j(a_2)$, using the inverse transformation, $m_j(a) = u^{-1}(EU_j(a; \hat{\gamma}); \hat{\gamma}), a \in A$, and the difference between them, Δm_j.
3. Start with $\varepsilon = 0$. Compute the optimal region using (10.8) and the aggregate DoR $\rho(\hat{\mu}, \varepsilon)$ using (10.5).
4. If $\rho(\hat{\mu}, \varepsilon) < \alpha$, increase ε by a small number $\Delta\varepsilon > 0$.
5. Repeat Step 5 until the aggregate DoR reaches the target level of α.[17]
6. Compute the AWC $\bar{\varepsilon}(\hat{\mu}, \alpha)$ using (10.6). Compute the RWC $\tilde{\varepsilon}(\hat{\mu}, \alpha)$ using (10.7).

10.3 Empirical Analysis

10.3.1 Data

We present the results for 218 adult Danes, a subsample of a larger field study by Harrison et al. (2018). The subjects for the original study were recruited from two internet-based panels with 165,000 active members combined. The sample frame consisted of 65,592 adult Danes between ages eighteen and seventy-five. The sample was stratified by sex and age across three regions of Denmark: greater Copenhagen, Jutland, and Funen and Zealand.[18] The completed sample consisted of 8,405 respondents, or 12.8% of the sample frame. Invitations were sent by email, and the subjects could participate in a survey using internet-browsers on their computers or mobile devices. The experiment

17. In practice, due to the discreteness of $\rho(\hat{\mu}, \varepsilon)$ it will often be impossible to match the target α exactly. We use linear interpolation to handle this issue.

18. Greater Copenhagen area was assigned a weight of 50%, and the other two regions were assigned equal weights of 25%.

was implemented as an artefactual field experiment with real monetary choices (Harrison and List 2004).

Slightly less than half of the sample were women and the average age was just less than fifty years. The majority of the sample had college education, and the distribution of income across different income brackets was roughly equal. Most of the participants were either employed as public servants or retired. More than 75% of our sample comes from the Greater Copenhagen area. Table D.2 in online appendix D provides a summary of the socio-demographic characteristics of our subsample.

The subjects made binary choices across sixty pairs of lotteries and answered a set of demographic questions. Once all the lottery choices were made, one of the choices was selected randomly for payoff. Table D.1 in online appendix D contains the battery of lotteries that were given to the subjects. This battery is based on designs by Loomes and Sugden (1998), Wakker et al. (1994), and Cox and Sadiraj (2008). Together they provide a powerful test of EUT and RDU.

10.3.2 Model Assumptions

We assume that the lotteries are compared according to their expected utilities (dropping an index for the round)

$$EU(a;\gamma) = \sum_{i=1}^{k} q_i(a)u(x_i(a);\gamma),$$

and the u function takes the constant relative-risk aversion form:

$$u(x;\gamma) = \frac{x^{1-\gamma}}{1-\gamma}.$$

We assume that the choice probability $p_j(a_2;\gamma,\mu)$ is given by the strong utility model in the logit form[19]

$$p_j(a_2;\gamma,\mu) = \Lambda\left(\frac{EU_j(a_2;\gamma) - EU_j(a_1;\gamma)}{\mu}\right),$$

where $\Lambda(\cdot)$ denotes the logistic cumulative density function, and $p_j(a_1;\gamma,\mu) = 1 - p_j(a_2;\gamma,\mu)$.

Nothing in our approach relies on assuming an EUT model or a strong utility model in the logit form. Online appendix C demonstrates several important

19. One could alternatively use certainty equivalent functions m instead of the aggregate utility functions U in the specification of the stochastic model, as in Bruhin et al. (2010) or Gaudecker et al. (2011). Using this alternative specification only changes the scale of the estimates of the noise parameter, and does not change the estimates of risk parameters or the estimated likelihoods of choosing each alternative. The algorithm for computing welfare costs and their magnitude would remain the same.

generalizations by regenerating all results assuming an RDU model of risk preferences, as well as assuming a different utility function, the expo-power utility function, and a different model of stochastic choice, the contextual utility model of Wilcox (2011).

We estimate the model separately for each subject to get the estimates of risk aversion and noise. We then use these estimates to compute the AWC and RWC for each subject according to the algorithm described above. Appendix 10.6 contains the details of the estimation procedure.

10.3.3 Welfare Costs

Figure 10.4 (Panel A) shows the distribution of AWC (in Danish kroner, DKK), and table 10.1 (Panel A) presents their summary statistics. The distribution of the AWC is composed of two clusters: a major cluster on the left end and a minor cluster on the right end of the support, so that overall the distribution is right-skewed. The major cluster is bell-shaped and fairly symmetric. The minor cluster appears to be bell-shaped but the number of observation in it is small.

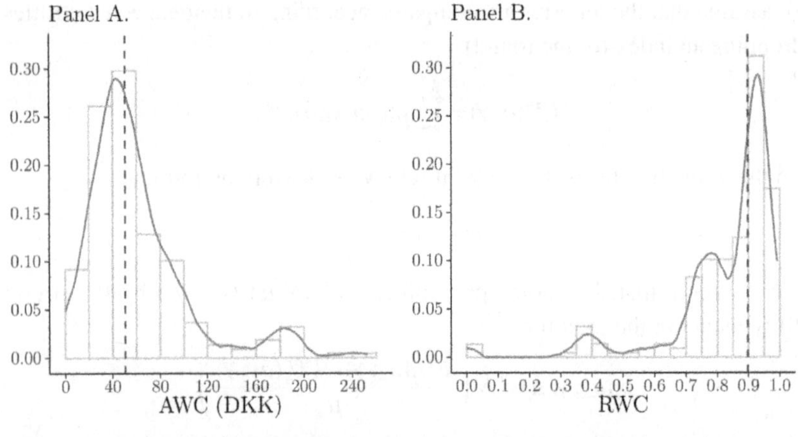

Note: The distributions of AWC (Panel A) and RWC (Panel B) for $\alpha = 0.95$. The bars are the histograms and the smooth lines are the kernel density estimates. The *y*-axis is scaled to show the proportion of observations falling into each bin. The dashed lines show the medians of the distributions. The AWC numbers are in DKK.

Figure 10.4: Distributions of AWC and RWC in the sample.

The AWC are, on average, modest in size. For $\alpha = 0.95$, the mean AWC are only 60.7 DKK (9.11 USD) and the median AWC are smaller, at 50.41 DKK (7.56 USD). There is substantial variation among subjects in their AWC. The smallest AWC are just 0 DKK (0 USD) while the maximum AWC are 252.52 DKK (37.88 USD), which is roughly four times as large as the mean AWC. The standard deviation of AWC at this level of DoR is 43.72 DKK (6.56 USD). For 50% of the subjects, the AWC lie within 33.23 DKK (4.98 USD) and 73.33 DKK (11 USD) .

Table 10.1
Summary statistics for AWC and RWC

α	Mean	SD	Min	Q1	Median	Q3	Max
Panel A. *AWC (DKK)*							
0.95	60.70	43.70	0	33.20	50.40	73.30	253.00
Panel B. *RWC*							
0.95	0.83	0.17	0	0.77	0.90	0.93	0.99

Notes: The table reports the summary statistics for the individual-level estimates of AWC and RWC computed at 0.95 DoR. The AWC numbers are in DKK. RWC are measured as proportions.

While the variation in the AWC between subjects is substantial, there is also considerable uncertainty at a subject level. Figure 10.5 (Panel A) shows the point estimates of the AWC (for $\alpha = 0.95$) for each subject along with the 95% confidence intervals around those estimates. The confidence intervals are computed using bootstrap methods. We rank subjects based on their point estimates of AWC. The vertical axis represents the percentile rank of each subject. The uncertainty in the estimates of AWC stems from the combined uncertainty in the estimates of risk aversion and noise and tends to increase with the percentile rank.

Figure 10.4 (Panel B) shows the distribution of RWC and table 10.1 (Panel B) presents their summary statistics. The distribution is mostly concentrated at the right end of the support and has a long left tail, resulting in a negative skew. The distribution features some observations on the left tail with unusually low RWC. For some of these outcomes, RWC are less than a half.

In contrast to the AWC, the RWC are high, which implies that although the AWC are modest in size, these costs represent a substantial portion of the monetary-equivalent welfare available in the choice environment. For $\alpha = 0.95$, the mean RWC are 0.83 and the median RWC are 0.9, so that around 90% of the relative welfare has to be sacrificed in order to rationalize this proportion

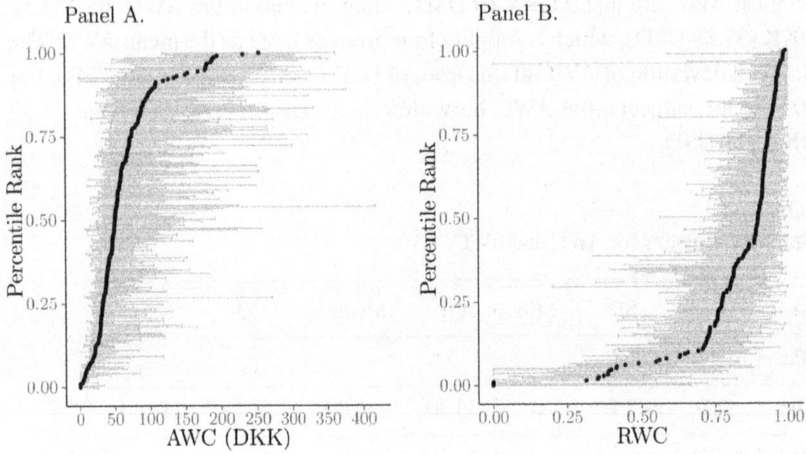

Note: Point estimates (in black) and confidence intervals (in gray) for AWC (Panel A) and RWC (Panel B) computed at $\alpha = 0.95$. Each point on the graph represents an estimate for a given subject. The estimates are ranked from lowest to highest, and the percentile rank of each subject is shown on the vertical axis. The horizontal error bars show the bootstrapped 95% confidence intervals around the point estimates.

Figure 10.5: Uncertainty in the individual-level estimates of AWC and RWC.

of choices. The RWC numbers show a modest variation across subjects. At $\alpha = 0.95$ the smallest amount of RWC is 0, while the maximum amount is 0.99, which is roughly 1.19 times as large as the mean amount. The standard deviation at this level of DoR is 0.17. For 50% of the subjects, the RWC lie within 0.77 and 0.93 at this level of DoR.

There is considerable uncertainty in the RWC at the individual level, just as we found for the AWC. Figure 10.5 (Panel B) shows the point estimates of RWC (at $\alpha = 0.95$) for each subject along with the 95% confidence interval around those estimates. Contrary to the AWC, however, the uncertainty in the RWC is highest at the lower percentile ranks. This uncertainty tends to decrease with the percentile rank.

The preceding analysis allow us to formulate the following result:

Result 1. *The welfare costs are low in terms of everyday economic activity, but are substantial for the choice environment in which they occurred.*

Comparing our results to Choi et al. (2014), we find that the subjects in our sample require a larger fraction of the total monetary-equivalent welfare to rationalize their choices. This difference can be explained by the differences in methods. The GARP-based measure used by Choi et al. is well known for its relatively mild requirements on choice consistency (Beatty and Crawford 2011).[20] For example, their primary measure does not even require choices to satisfy first-order stochastic dominance. It is far from clear that we should expect the QIS to track a RP if an assumption of stochastic dominance is systematically rejected. We would then lack the crucial basis for believing that we are observing an instance of agency.

10.3.4 Marginal Welfare Costs

So far we have looked at the distributions of AWC and RWC for only one level of the target DoR. This analysis does not tell us how quickly welfare costs grow as the rationalizability requirements become tighter, and in general what the shape of the costs as functions of α is. Figure 10.6 provides an answer to these questions by showing the median welfare costs as functions of α across all the subjects in the sample, with the dashed lines corresponding to the 5% and 95% empirical quantiles. The lowest possible target DoR in our context is 0.5, since the choice is binary. However, the welfare costs stay at 0 for the median subject until α crosses the 0.61 mark.

Panel A on figure 10.6 shows the graphs of the AWC in relation to α. The median AWC tend to be a convex function of α: at first increasing the DoR requires relatively little AWC, but as the target becomes higher, each additional percentage point of DoR costs more and more in terms of AWC. The graph for the RWC in Panel B of figure 10.6 is in a sense the mirror image of the AWC. The RWC tends to be a concave function of α. For small values of DoR extra percentage points of change require high welfare costs, but as the target increases these extra points become less costly in relative terms. These observations allow us to formulate the next result:

Result 2. *The marginal absolute (relative) welfare costs are increasing (decreasing) with the increase in the target DoR.*

This result can be explained by the way our measures of welfare costs are computed. Starting from a given default DoR, $\rho(\hat{\mu}, 0)$, we gradually increase ε until the DoR reaches the target, $\rho(\hat{\mu}, \varepsilon) = \alpha$. The low marginal AWC at low α targets imply there are many choices that can be easily rationalized by small ε, since the difference in the certainty equivalents between the alternatives must be low. At high target DoR more choices have to be rationalized, but no

20. Another explanation could be that there are, of course, systematic differences in the samples used. The Choi et al. (2014) experiment was conducted in the Netherlands, while our experiment was conducted in Denmark. We believe this explanation to be unlikely a priori. Blow et al. (2008), who employ a revealed preference approach similar to Choi et al. (2014), show that their Danish sample also has a high degree of choice consistency, thus pointing to the difference in methods rather than the difference in samples as the likely explanation for why results differ.

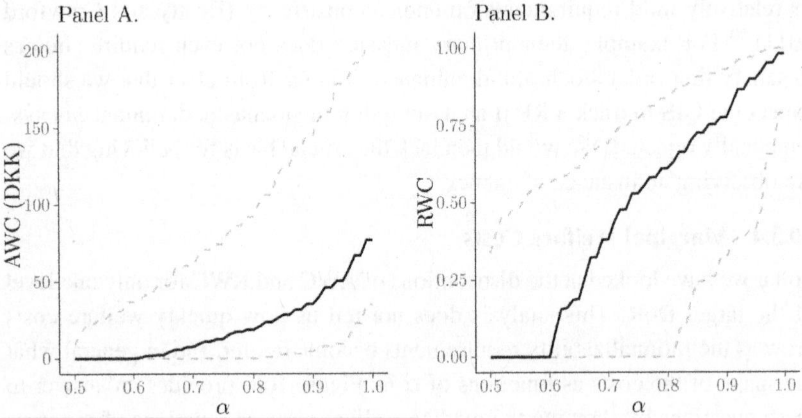

Note: The AWC (Panel A) and the RWC (Panel B) as functions of the target DoR (α). The black solid lines are the median welfare costs for each level of α. The dashed lines below (above) the solid line represent the 5% (95%) quantiles of the welfare costs.

Figure 10.6: Welfare costs as functions of the Target DoR.

"easily rationalizable" choices are left. Increasing DoR requires tapping into choice pairs with higher differences in certainty equivalents, and hence higher marginal AWC. The implications for the RWC graphs are the converse. At low target DoR the marginal RWC are high, since rationalizing many choices with small differences in certainty equivalents requires the whole difference. At high targets fewer such choice pairs remain and the marginal RWC decrease.

10.3.5 Relationship Between the Measures

We now consider the relationship between the two welfare costs measures and the default DoR (DDoR). We ask whether people with lower AWC also have lower RWC, and formally test the previous observation that people with lower DDoR tend to have higher costs. The motivation behind these questions is that it is intuitive to expect a positive relation between AWC and RWC. It does not follow, however, directly from the method of their computation. Only if preferences are held constant must higher AWC imply higher RWC, but there is no such prediction when preferences are not the same when making comparisons across subjects. Likewise, even though it is natural to expect that people with higher DDoR have lower costs, we do not have a formal basis to expect this observation to hold a priori.

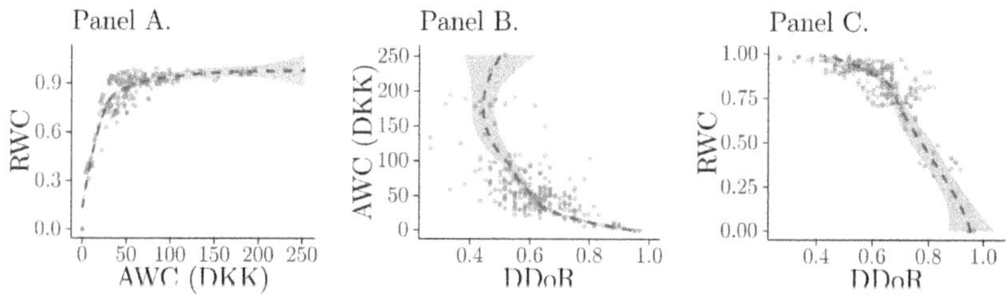

Note: Scatterplots between three pairs of measures: AWC, RWC, and DDoR. The welfare costs are evaluated at $\alpha = 0.95$. The dots represent individual subjects. The dashed lines are the smooth fitted lines estimated using local polynomial regressions, and the shaded regions are the estimated 95% confidence intervals.

Figure 10.7: Relationship between AWC, RWC, and DDoR.

Figure 10.7 (Panel A) shows a scatterplot of the RWC (y-axis) against the AWC (x-axis) computed at $\alpha = 0.95$.[21] A clear positive association between the two measures can be observed. The Kendall rank correlation between the two measures is 0.54 (the ranking of subjects by the two measures is the same 77% of the time) and is significantly positive, with p-value < 0.001. The relation between RWC and AWC is nonlinear, and has a concave shape.

Figure 10.7 (Panel B) shows the scatterplot of the AWC (y-axis) against the DDoR (x-axis), and confirms our earlier observation from the analysis of marginal costs. There is a moderate negative association between the DDoR and the AWC. The Kendall rank correlation between the two measures is –0.47 (the ranking of subjects by the two measures is the opposite 74% of the time) and is significantly negative, with p-value < 0.001. The relation between them is again nonlinear, and has a convex shape.

Figure 10.7 (Panel C) shows the scatterplot of the RWC (y-axis) against the DDoR (x-axis). We can immediately see a very tight negative association between the two measures. The Kendall rank correlation between them is –0.56 (the ranking of subjects by the two measures is the opposite 79% of the time) and is significantly negative, with p-value < 0.001. The relation is again slightly nonlinear and has a concave shape.

These observations allow us to formulate the following result:

Result 3. *People with higher absolute welfare costs tend to have higher relative welfare costs. People with higher default degree of rationalizability tend to have lower absolute welfare costs and relative welfare costs.*

21. The results in this section remain quantitatively similar if we use $\alpha = 0.9$ or 0.99.

This result implies that there is a certain degree of consistency between the measures we introduce. Moreover, this consistency works in the way we expect. This property could not have been deduced from the method by which these measures are constructed. It is possible that a subject with high AWC has preferences such that the differences in the certainty equivalents are even higher, and the RWC are actually low. We do, in fact, observe such individual cases. But the general tendency is for the subjects to have the same ordering, whether it is measured according to the absolute or relative measure of welfare costs.

There is also a negative relationship between the DDoR and the welfare that the two costs measure. This implies that people who make more consistent choices, measured by the DDoR, also require less welfare costs to rationalize their choices. This is also an intuitive property, but it is hard to see a priori why it should hold even though the data indicate that it does, with the relation between the default DoR and the RWC being particularly strong. The relative strength of this relationship, compared to the relationship with the AWC, can be partially attributed to the fact that both the default DoR and the RWC are relative measures defined on the unit interval. Nonetheless, such a strong relationship is remarkable, given that the two measures address different questions.

10.3.6 Welfare Costs and Noise

Our approach is largely motivated by the desire to attach a *normative* economic meaning to the noise parameter. It is therefore of interest to look at the relationship between the two welfare cost measures we introduced and noise, as well as the relationship between the DDoR and noise. Higher noise does translate into higher welfare costs if preferences are kept constant, but no prediction is available for comparisons between subjects. It is natural to expect, however, that this property should also hold between subjects. Given the negative association between the DDoR and the cost measures, it is also natural to expect that higher noise translates into lower default DoR, but whether it does is an empirical question.

Figure 10.8 shows the scatterplots of (from left to right) the AWC and RWC and the DDoR (on the y-axis) against the logarithm of noise (x-axis).[22] The three panels confirm our hypotheses. We do see that higher noise is associated with higher AWC and RWC and lower DDoR, although the strength of

22. We truncate the logarithm noise at 15, in order to make the graph more readable. This excludes one subject.

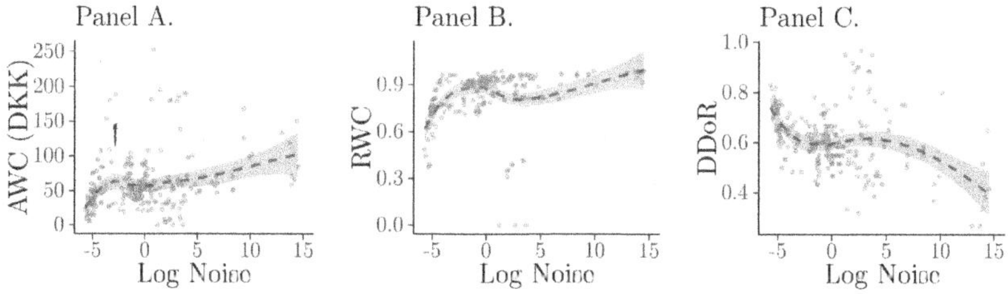

Note: Scatterplots between three measures (AWC, RWC, DDoR) and (the logarithm of) noise. The welfare costs are evaluated at $\alpha = 0.95$. The dots represent individual subjects. The dashed lines are the smooth fitted lines estimated using local polynomial regressions, and the shaded regions are the estimated 95% confidence intervals. The graphs drop one subject with log noise higher than 15.

Figure 10.8: Relationship between welfare costs and noise.

this association differs across the measures. It is small, though statistically significant, for the AWC.[23] The association with the RWC is much stronger.[24] Finally, the association with the DDoR (in absolute terms) is slightly weaker than the association with the RWC, but much stronger than the association with the AWC.[25]

A notable feature in these results, most pronounced in the relationship between noise and the DDoR and the RWC, is that there is an outer boundary that constrains the values. On Panel B of figure 10.8 this boundary constrains the values of the RWC from above, and on Panel C of figure 10.8 this boundary constrains the values of the DDoR from below. This pattern suggests that for given noise the RWC (DDoR) cannot be higher (lower) than a certain value, defined by this boundary.

These findings lead us to the next result:

Result 4. *People with higher noise tend to have higher absolute and relative welfare costs and lower default degree of rationalizability. For any given value of noise there appears to exist a maximum (minimum) amount of absolute and welfare costs (degree of rationalizability) that one can have.*

23. The Kendall rank correlation between the two measures is 0.18 with p-value < 0.001 (the ranking of subjects by the two measures is the same 59% of the time). The weakness of the association can be seen by the substantial variation in the AWC at the high values of noise, which means that there are many subjects with high estimates of noise but low AWC.

24. The Kendall rank correlation between the two measures is 0.39 with p-value < 0.001 (the ranking of subjects by the two measures is the same 69% of the time).

25. The Kendall rank correlation between the two measures is –0.28 with p-value < 0.001 (the ranking of subjects by the two measures is the opposite 64% of the time).

The first part of this result confirms our intuitive guesses. We do see some association between noise and welfare costs, which implies that noise contains some information about welfare costs and choice consistency, but this information is imprecise. Despite big differences in noise estimates between some subjects, their RWC need not be that different. Similarly, some subjects might appear to have high welfare costs based on the noise measure, while in fact their AWC are not nearly as large.

The second part of the result is unexpected and remarkable. It says that there is a regularity in the relation between noise, welfare costs and default DoR. This regularity is in the form of a boundary that constraints the possible values.

10.3.7 Socio-Demographic Covariates of Welfare Costs

The estimates of welfare costs vary substantially between subjects in our sample. Here we attribute this variability to the observable socio-demographic characteristics of the subjects. We focus on sex, age, education, work, income, housing, family, and health characteristics. Online appendix E contains the details of the descriptive estimation.

Women tend to have higher AWC than men, although the effect is not statistically significant. The RWC are not statistically or economically different between men and women. Welfare costs tend to be higher for older subjects. The AWC tend to increase monotonically with the age group, but the effect is not precisely estimated. The RWC are higher for subjects older than thirty years than for younger subjects, but there is no statistically significant difference between the three age groups above thirty years. College education has a beneficial impact on the welfare costs relative to vocational training. The effect is most pronounced for the RWC and subjects with five or more years of college, although it is not precisely estimated. Subjects who are employed as public servants tend to have lower AWC and RWC than subjects employed as students or workers. The effect of income is mixed and not precisely estimated. Subjects with medium and medium-high levels of income tend to have higher welfare costs, while subjects with very high levels of income tend to have lower welfare costs. The type of housing a subject occupies and the type of ownership does not appear to have a meaningful impact on welfare costs. Similarly, the effects of civil or parenting status are small and not statistically significant. Subjects show some systematic variation by their health status and life experiences with the effects most pronounced for RWC. For instance, subjects who experienced death of a household intimate tend have higher RWC.

These observations lead us to the following result:

Result 5. *Having higher welfare costs is associated with higher age. The RWC are not significantly different for men and women, although the AWC for women tend to be higher.*

10.4 Generalizing the QIS for Behavioral Welfare Economics

Our construction of the QIS to identify real choice patterns that can anchor welfare assessments extends, in a quantitative framework, the normative elements of Dennett's IS. To complete our argument that this provides the basis for a general philosophy of economics—which in turns licenses interpreting the ontology of the discipline as RPs—we must show that we can do more than interpret some selected sets of choice data. In addition, we must demonstrate that we can use the QIS to integrate leading recent strands of theory in behavioral welfare economics.

We begin with a reflection on why such integration is important. Descriptive (positive) behavioral economics has recurrently complicated models of the consistency of observed choices by people, at least outside of tightly institutionally regulated settings. Faced with these findings, behavioral welfare economists cannot go on effortlessly assuming that isolated observed choices are straightforward indicators of preferences that policy advisors should help agents satisfy as efficiently as possible. Methods must be found for consistently extracting real preference patterns from noisy choice data.

We pointed out that the most general relationship between economic theory and Dennett's original version of the IS is that RPT, and more specifically GARP, anticipate Dennett's understanding of the basis of belief and preference attribution. Now we can add that our specific measures described in this chapter are similar to the Critical Cost Efficiency Index (CCEI) of Afriat (1972) used to evaluate the degree of consistency of a body of choices with GARP. Just like our relative cost measure, CCEI is defined on the unit interval, and its complement shows what proportion of monetary value an agent can waste without losing rationalizability of her choices by some utility function. While GARP, like the IS, provides a basis for qualitative assessment, we put more structure on it that more directly ties it to quantitative evidence.

This point is crucial for linking the QIS to the theory of real patterns. The commonly cited information-theoretic refinement of Dennett's concept given by Ladyman and Ross (2007, 233) has two parts. The first requirement is that the pattern must be projectible out of sample. As formally proven by Reny (2015), satisfaction of this requirement for any set of choices is guaranteed by GARP. But this falls short of the second requirement for pattern reality, that this projectability not be decomposable, holding information constant, into a complete basis in more granular patterns. To show that this applies to GARP, one must rely on a model that does not itself assume GARP. Polisson and Renou (2016) show that, if an agent's budget set is linear, then rationalizability of a finite choice set by a concave utility function depends on the nonuniqueness of the inequalities used to establish GARP. The restriction to nonlinear budget sets is severe where economic significance is concerned, but the existence of cases where the pattern cannot be decomposed without adding information is sufficient to satisfy the second part of the Ladyman and Ross analysis.

For empirical scientists, however, satisfaction of purely formal conditions on the reality of a hypothesized pattern serves merely as a check on technical consistency. It cannot, as the discussion above illustrates, assure us that the pattern in question is significant for projecting actually observed cases. We now comment on recent approaches in behavioral welfare economics from this perspective.

Our measures of the welfare costs of noisy behavior are consistent with the model-based approach advocated by Manzini and Mariotti (2014). That is, in order to calculate our welfare cost measures, we assume specific deterministic and stochastic models of a choice-generating process. (We avoid the more familiar semantics of "decision-making" here to distance ourselves, in keeping with the IS, from reductive internalism about preferences and beliefs.) These assumptions allow us to derive precise values of welfare costs. We recognize the potential sensitivity of our results to these assumptions, and address them with some robustness checks in online appendix C.

Our absolute and relative welfare cost measures allow for conveniently evaluation of the economic significance of choice randomness and its relative magnitude, and to compare magnitudes of choice randomness across people. The implications of these measures for an agent's behavior, however, will ultimately depend on the underlying model of the source of choice randomness adopted. This is an important point, since different "source" models of stochastic choice often lead to the same choice likelihoods, such as the likelihoods generated by the multinomial logit model presented above. Are such models interpretative variants of RPs in economic choice?

For instance, the random utility model due to Marschak (1960) assumes that when an agent makes an optimal choice, the choice randomness is due to idiosyncratic perturbations in her utility function that are unobservable to a researcher. The noise parameter in the random utility model is then proportional to the variance of the unobserved component of utility. High estimated welfare costs would imply that the stochastic part of the structural model dominates the deterministic part, such that the structural model cannot descriptively explain the agent's choices well. The welfare costs can then be viewed as measures of a model's fit.[26]

Recent studies offer an alternative view on choice randomness as an optimal response to costly frictions in a choice-generating process. For example, these frictions may be caused by the need to collect the relevant information to make a choice, as in the rational inattention models of Caplin and Dean (2015) and Matêjka and McKay (2015). The noise parameter in a rational inattention model represents marginal information costs. The estimates of welfare costs in this type of model can then be interpreted as aggregate information costs, or losses that an agent incurs relative to an ideal case of no information costs.

Another view on choice randomness arises in the pursuit of multiple goals that cannot be obtained simultaneously (Swait and Marley 2013; Wallin et al. 2018). An agent is assumed to balance the goal of choosing the best available alternative with the goal of having diversity in choices. Noise parameters in this model represent the relative weight of the second goal. The estimates of welfare costs in models of this type can be interpreted as the economic value that an agent places on the goal of preserving diversity or, alternatively, as the loss an agent incurs relative to a case of having a single goal of choosing the best alternative.

The measuring procedure we have described in this chapter abstracts from all of these specific mechanistic interpretations of choice and randomness, suggesting that they should all be regarded as possible types of instances of RPs best identified at the more general level of the QIS.

The intuition behind the computation of our measures also links it to a literature on payoff dominance in experiments (Harrison 1994, 1992; Harrison and Morgan 1990; Harrison 1989). This literature shows that allowing for small deviations from optimal behavior, just as we do, allows for rationalization of supposedly anomalous choice behavior observed in experimental studies.

Unsurprisingly, the method we have described and generated is also complementary to work by some of the present authors that preceded explicit articulation of the QIS. Harrison and Jia Min Ng (2016) and Harrison and J. M. Ng (2018), Harrison et al. (2022) and Gao et al. (2023) anticipate the QIS when they evaluate losses of consumer surplus resulting from suboptimal insurance choices. Harrison and Ross (2018) apply the same approach to evaluate suboptimal portfolio investments. Their measure of lost consumer surplus is similar to our AWC measure, with both being based on computing certainty equivalents. A key difference noted previously, however, is that these

26. Halevy et al. (2018) provide an example of how welfare costs can be used as a measure of fit for parametric models of risk preferences.

studies use two experimental tasks: one for preference estimation and the other for welfare evaluation, whereas here we estimate welfare costs resulting from stochastic choice in the same task as was used to reveal the relevant preferences. The approach that we take in this study does not rely on an independent risk metric, but rather relies on a specific noise structure to "bootstrap" a measure of welfare costs. This should not be taken to imply that we think that such bootstrapping is always the better methodology; parsimony of evidential bases can trade off against other methodological virtues, such as precision of estimates, transparency of tasks to subjects, or even professional conventions when determining priors for normative evaluation.

Our approach can be related to studies that estimate structural models of choice under risk and over time. Andersen et al. (2008) use the strict utility model to structurally estimate risk and time preferences of a representative sample of the Danish population, and note that noise estimates are higher in the risk task than in the discounting task. Gaudecker et al. (2011) use a representative sample of the Dutch population to estimate subjects' risk preferences using a model of stochastic choice that is a hybrid between the multinomial logit and tremble models, and features two measures of choice randomness due to noise: a Fechner parameter that is common to all subjects, and a "tremble" parameter that is specific to each subject.[27] While these studies typically focus on estimates of risk and time preferences, and do not interpret the estimates of the stochastic component, we explicitly focus on the estimates of the stochastic part and provide a systematic approach to economically interpret the estimates of choice randomness. Bland (2020) considers mixture specification over pooled choices, contrasting one "baseline" model, such as EUT, as one of the data generating processes (DGP) with an "alternative" model, such as prospect theory, as the other DGP. He then calculates the CE of choices using the deterministic core of the "rational" model DGP, thereby evaluating potential welfare losses from using a "behavioral" DGP, as well as from noise for both DGPs. We reject the idea that rationalizability should be used to distinguish "rational" from "non-rational" choice, regarding the discrimination of agency from non-agency as the limiting ambition of the welfare economist where such philosophical interpretation is concerned (Harrison and Ross 2023). But the general logic of allowing the estimated structural model of noise to provide a basis for welfare evaluation is consistent with our approach.

27. Their model also allows for random coefficients on core parameters for risk preferences. Although this specification allows for stochastic variation in those parameters, it is conventionally interpreted as reflecting unobserved heterogeneity in these parameters across subjects, "unobservable" only in the sense that the researcher cannot account for the variation with observable characteristics of the subject or task.

We provide economic measures of choice randomness (or consistency), which link to studies on the quality of choice. Choi et al. (2007) study choice under risk in a laboratory experiment in which they present subjects with convex budget sets for two Arrow securities. This design allows them to gauge subjects' choice quality using a measure of GARP-consistency, a standard technique in the revealed preference approach to consumer demand. They find that subjects' behavior is highly consistent with GARP. Choi et al. (2014) expand the analysis by using a representative panel of the Dutch population. They also find a high degree of GARP-consistency in risky choices, which varies, however, with education, sex, and age. Beatty and Crawford (2011) show that while behavior in a wide range of situations is highly GARP-consistent, this might be a result of a misspecified measure of consistency. They propose an alternative to the traditional CCEI measure, which is based on predictive success, and show that the CCEI measures of GARP-consistency are overinflated, and hence that the actual consistency of choices is much lower. Hey (2001) studies quality of choice under risk in a laboratory experiment, and asks whether choice consistency, which he defines as the number of choices that remain the same between task repetitions, improves with experience. He finds mixed evidence of a positive effect of experience on choice consistency. We rely on a parametric measure of choice consistency and find a lower degree of consistency than in the studies that use the non-parametric revealed preference approach.

Finally, our approach is also related to recent literature on rational inattention. Matêjka and McKay (2015) show that when an agent faces information costs, optimal behavior is stochastic choice,[28] and that under certain conditions the resulting choice likelihoods are represented by the multinomial logit specification. Cheremukhin et al. (2015) apply a model of rational inattention to risky choices and estimate the shape of the cost-of-information function[29] in a laboratory experiment with student subjects. Caplin and Dean (2015) develop a revealed preference test of rational inattention theories with general cost-of-information functions. The noise parameter in the rational inattention models has the interpretation of marginal information costs. Since our method converts the noise parameter into monetary percentage terms, we can alternatively

28. In the absence of information costs, an agent would collect all the available information about and thus know with certainty which option is the best. The introduction of information costs makes collecting all available information suboptimal. Without full information, the agent will remain uncertain about which option is the best.

29. The cost-of-information function describes the relationship between the amount of information an agent collects and the cognitive costs involved in processing that information.

interpret our estimates as the monetary value of these marginal information costs.

10.5 Conclusion

Stochastic choice has become an increasingly central area of both theoretical and empirical research in both descriptive and normative economics. While the existing literature mainly focuses on the sources of choice randomness, its economic consequences are less well understood, and hence it raises questions about the justification for regarding behavioral models as providing a basis for welfare assessment. In this chapter we respond to this challenge by quantitatively interpreting Dennett's IS, and capturing its normative element by seeking to measure absolute and relative costs of extracting RPs from noise. We develop tools to assess the economic significance of noise and apply them to a sample from the general Danish population in an artefactual field experiment.

We introduce three interconnected concepts: imperfect rationalizability, optimal region, and degree of rationalizability. Fixing the degree of rationalizability at a certain target level, we vary the amount of imperfect rationalizability, which in turn affects the optimal region, to make the proportion of subjects' choices falling in the optimal region equal the target level. This measured degree of imperfect rationalizability represents the welfare costs, or monetary welfare allowed to be wasted, that is required to rationalize a given proportion of choices. The resulting welfare costs can be expressed both in absolute monetary terms against prices of certainty equivalents, and in relative (to the actual stakes of the choice environment) terms.

We compute the absolute welfare costs and relative welfare costs at the individual level in an experiment with binary-choice lotteries. This application provides an example of the kind of economically substantial empirical findings our approach can support. Some of these coincide with previous findings, and some of are new. We find that the AWC are not economically significant in our sample, while the RWC are economically significant. In other words, the welfare costs are tiny if viewed from a broad perspective of economic activity, but they are substantial if viewed from the perspective of this particular choice experiment. As compared to Choi et al. (2014), who employ a relative measure based on the consistency with GARP, our estimates of RWC are much larger. We attribute the difference in results to the difference in the methods, with our method imposing stricter requirements that take the demands of agency more seriously. The ultimate justification for this stringency is that agency—the robust sensitivity of a system's behavior to contingencies that affect its welfare—is an RP, which is identified when certain criteria, such as stochastic dominance, are in place, and not otherwise.

Since our welfare cost measures depend on the target level of rationalizability α, we study the shape of the relation between α and these welfare costs. We find that the AWC increase in α at an increasing speed, while the RWC increase in α at a decreasing speed. The difference in these two relations is explained by the way our method of computation works. Subjects with higher AWC tend to have higher RWC. Also, a lower DDoR is associated with higher AWC and RWC: Subjects who start out with low default degree of rationalizability require a higher cost to reach a given degree of rationalizability. Looking at the relationship between our cost measures and raw estimates of noise reveals that they are positively associated, though our measures do not have such a wide range, which allows for sensible comparisons across subjects and allows us to make judgments about the magnitudes of choice inconsistencies.

The articulation of the IS transformed cognitive science, providing an understanding of beliefs, preferences and their cognate concepts that unifies study of the wide range of intentional systems, including people, other animals, composite agents, and prospective artificial agents (Brook and Ross 2002). The IS could not have exerted this influence had behavioral scientists not been persuaded that intentional characterizations are more than loose attempts to infer processes in nervous systems that "realize" them. The argument that they are RPs supplies the basis for this persuasion.

Similarly, behavioral economics casts doubt on the cogency of welfare economics by identifying choices with processes that lack the normative aspects *traditionally* associated with preferences, and which also arise for beliefs about subjective probabilities. In response we can borrow from the history of cognitive science by quantizing the IS—that is, expressing its qualitative discourse frame in quantitative terms—and we can escape the concern that in doing so we are merely playing technical tricks by showing empirically that the patterns we can consequently track are real.

10.6 Acknowledgments

We are grateful to the Danish Social Science Research Council (Project #12-130950) for financial support. We thank Harold Kincaid for his critical suggestions.

Appendices

A. Estimation Procedure

Computation of the welfare costs relies on structural estimates of risk preferences γ and noise μ. We implement the estimation in the standard fashion by maximizing the Bernoulli log-likelihood function at the level of a subject:

$$(\hat{\gamma}, \hat{\mu}) = \arg\max_{\gamma,\mu} \sum_{j=1}^{n} \left(y_j \ln p_j(a_2; \gamma, \mu) + (1 - y_j) \ln p_j(a_1; \gamma, \mu) \right),$$

where $y_j \equiv \mathbb{I}(c = a_2)_j$ is an indicator variable that takes a value of 1 whenever an alternative a_2 is chosen in round j. The alternative a_2 is taken to be the one on the right side of the screen without loss of generality, and we no longer assume that it gives the highest aggregate utility in all the rounds.

B. Proofs

Consider an implicit function $\rho(\mu, \varepsilon) = \alpha$. From the implicit function theorem, it follows that

$$\frac{d\varepsilon}{d\mu} = -\frac{\partial\rho/\partial\mu}{\partial\rho/\partial\varepsilon}.$$

The denominator of this expression is

$$\frac{\partial\rho}{\partial\varepsilon} = \frac{\partial}{\partial\varepsilon} \int_{a_l^*(\varepsilon)}^{a_h^*(\varepsilon)} p(a)da = p(a_h^*(\varepsilon))a_h^{*'}(\varepsilon) - p(a_l^*(\varepsilon))a_l^{*'}(\varepsilon) \geqslant 0,$$

since $a_h^{*'}(\varepsilon) \geqslant 0$, and $a_l^{*'}(\varepsilon) \leqslant 0$.

In order to show the sign of the numerator, we restrict our attention to the binary choice case, since it is the setting of our primary interest. Recall that

$$p(a_2; \gamma, \mu) = \Lambda\left(\frac{U(a_2; \gamma) - U(a_1; \gamma)}{\mu}\right).$$

Then

$$\frac{\partial p(a_2; \gamma, \mu)}{\partial\mu} = \Lambda'\left(\frac{U(a_2; \gamma) - U(a_1; \gamma)}{\mu}\right)(U(a_2; \gamma) - U(a_1; \gamma))(-\mu^2) < 0,$$

since alternative a_2 gives the highest certainty equivalent by our assumption. Therefore,

$$\frac{\partial\rho}{\partial\mu} = \begin{cases} \frac{\partial p(a_2;\gamma,\mu)}{\partial\mu}, & \varepsilon < \Delta m, \\ 0, & \varepsilon \geqslant \Delta m, \end{cases}$$

so that $\partial\rho/\partial\mu \leqslant 0$. Together the two results imply that $d\varepsilon/d\mu \geqslant 0$.

References

Afriat, S. N. 1967. "The Construction of Utility Functions from Expenditure Data." *International Economic Review* 8 (1): 67–77.

Afriat, S. N. 1972. "Efficiency Estimation of Production Function." *International Economic Review* 13 (3): 568–598.

Agranov, M., and P. Ortoleva. 2017. "Stochastic Choice and Preferences for Randomization." *Journal of Political Economy* 125 (1): 40–68.

Alekseev, A., G. W. Harrison, M. I. Lau, and D. Ross. 2024. *Deciphering the Noise: Real Patterns in Welfare from Incentivized Choice.* CEAR Working Paper 2024-03. Center for the Economic Analysis of Risk, Robinson College of Business, Georgia State University.

Andersen, S., G. W. Harrison, M. I. Lau, and E. E. Rutström. 2008. "Eliciting Risk and Time Preferences." *Econometrica* 76 (3): 583–618.

Ballinger, T. P., and N. T. Wilcox. 1997. "Decisions, Error and Heterogeneity." *Economic Journal* 107 (443): 1090–1105.

Beatty, Timothy K. M., and I. A. Crawford. 2011. "How Demanding Is the Revealed Preference Approach to Demand?" *American Economic Review* 101 (6): 2782–2795.

Becker, G. S. 1962. "Irrational Behavior and Economic Theory." *Journal of Political Economy* 70 (1): 1–13.

Bland, J. R. 2020. "Measuring and Comparing Two Kinds of Rationalizable Opportunity Cost in Mixture Models." *Games* 11 (1).

Blow, L., M. Browning, and I. Crawford. 2008. "Revealed Preference Analysis of Characteristics Models." *Review of Economic Studies* 75 (2): 371–389.

Bone, J., J. Hey, and J. Suckling. 1999. "Are Groups More (Or Less) Consistent Than Individuals?" *Journal of Risk and Uncertainty* 18 (1): 63–81.

Brook, A., and D. Ross. 2002. "Dennett's Position in the Intellectual World." In *Daniel Dennett,* edited by A. Brook and D. Ross, 3–40. Cambridge University Press.

Bruhin, A., H. Fehr-Duda, and T. Epper. 2010. "Risk and Rationality: Uncovering Heterogeneity in Probability Distortion." *Econometrica* 78 (4): 1375–1412.

Camerer, C. F. 1989. "An Experimental Test of Several Generalized Utility Theories." *Journal of Risk and Uncertainty* 2 (1): 61–104.

Caplin, A., and M. Dean. 2015. "Revealed Preference, Rational Inattention, and Costly Information Acquisition." *American Economic Review* 105 (7): 2183–2203.

Carbone, E., and J. D. Hey. 2000. "Which Error Story Is Best?" *Journal of Risk and Uncertainty* 20 (2): 161–176.

Chambers, C. P., and F. Echenique. 2016. *Revealed Preference Theory.* Cambridge University Press.

Cheremukhin, A., A. Popova, and A. Tutino. 2015. "A Theory of Discrete Choice with Information Costs." *Journal of Economic Behavior & Organization* 113 (May): 34–50.

Choi, S., R. Fisman, D. Gale, and S. Kariv. 2007. "Consistency and Heterogeneity of Individual Behavior Under Uncertainty." *American Economic Review* 97 (5): 1921–1938.

Choi, S., S. Kariv, W. Müller, and D. Silverman. 2014. "Who Is (More) Rational?" *American Economic Review* 104 (6): 1518–1550.

Cox, J. C., and V. Sadiraj. 2008. "Risky Decisions in the Large and in the Small: Theory and Experiment." In *Risk Aversion in Experiments,* edited by J. C. Cox and G. W. Harrison, 12:9–40. Research in Experimental Economics. Emerald Group Publishing Limited.

Dennett, D. C. 1989. *The Intentional Stance.* MIT Press.

Dennett, D. C. 1991. "Real Patterns." *Journal of Philosophy* 88 (1): 27–51.

Gao, Xiaoxue Sherry, G. W. Harrison, and R. Tchernis. 2023. "Behavioral Welfare Economics and Risk Preferences: A Bayesian Approach." *Experimental Economics* 26 (2): 273–303.

Gaudecker, Hans-Martin von, A. van Soest, and E. Wengstrom. 2011. "Heterogeneity in Risky Choice Behavior in a Broad Population." *American Economic Review* 101 (2): 664–694.

Gneezy, U., and J. Potters. 1997. "An Experiment on Risk Taking and Evaluation Periods." *Quarterly Journal of Economics* 112 (2): 631–645.

Gul, F., and W. Pesendorfer. 2006. "Random Expected Utility." *Econometrica* 74 (1): 121–146.

Halevy, Y., D. Persitz, and L. Zrill. 2018. "Parametric Recoverability of Preferences." *Journal of Political Economy* 126 (4): 1558–1593.

Harless, D. W., and C. F. Camerer. 1994. "The Predictive Utility of Generalized Expected Utility Theories." *Econometrica* 62 (6): 1251–1289.

Harrison, G. W. 1989. "Theory and Misbehavior of First-Price Auctions." *American Economic Review* 79 (4): 749–762.

Harrison, G. W. 1992. "Theory and Misbehavior of First-Price Auctions: Reply." *American Economic Review* 82 (5): 1426–1443.

Harrison, G. W. 1994. "Expected Utility Theory and the Experimentalists." *Empirical Economics* 19:223–253.

Harrison, G. W., Lasse J. Jessen, M. Lau, and D. Ross. 2018. "Disordered Gambling Prevalence: Methodological Innovations in a General Danish Population Survey." *Journal of Gambling Studies* 34 (1): 225–253.

Harrison, G. W., and J. A. List. 2004. "Field Experiments." *Journal of Economic Literature* 42 (4): 1009–1055.

Harrison, G. W., and P. Morgan. 1990. "Search Intensity in Experiments." *Economic Journal* 100 (401): 478–486.

Harrison, G. W., K. Morsink, and M. Schneider. 2022. "Literacy and the Quality of Index Insurance Decisions." *The Geneva Risk and Insurance Review* 47 (1): 66–97.

Harrison, G. W., and J. M. Ng. 2018. "Welfare Effects of Insurance Contract Non-Performance." *The Geneva Risk and Insurance Review* 43 (1): 39–76.

Harrison, G. W., and Jia Min Ng. 2016. "Evaluating the Expected Welfare Gain from Insurance." *Journal of Risk and Insurance* 83 (1): 91–120.

Harrison, G. W., and D. Ross. 2018. "Varieties of Paternalism and the Heterogeneity of Utility Structures." *Journal of Economic Methodology* 25 (1): 42–67.

Harrison, G. W., and D. Ross. 2023. "Behavioral Welfare Economics and the Quantitative Intentional Stance." In *Models of Risk Preferences: Descriptive and Normative Challenges,* edited by G. W. Harrison and D. Ross, 22:7–67. Research in Experimental Economics. Emerald Publishing. ISBN: 978-1-83797-269-2.

Hey, J. D. 2001. "Does Repetition Improve Consistency?" *Experimental Economics* 4 (1): 5–54.

Hildenbrand, W. 1994. *Market Demand: Theory and Empirical Evidence.* Princeton University Press.

Holt, C. A., and S. K. Laury. 2002. "Risk Aversion and Incentive Effects." *American Economic Review* 92 (5): 1644–1655.

Infante, G., G. Lecouteux, and R. Sugden. 2016. "Preference Purification and the Inner Rational Agent: A Critique of the Conventional Wisdom of Behavioural Welfare Economics." *Journal of Economic Methodology* 23 (1): 1–25.

Ladyman, J., and D. Ross. 2007. *Every Thing Must Go.* Oxford University Press.

Loomes, G., and R. Sugden. 1995. "Incorporating a Stochastic Element into Decision Theories." *European Economic Review* 39 (3): 641–648.

Loomes, G., and R. Sugden. 1998. "Testing Different Stochastic Specifications of Risky Choice." *Economica* 65 (260): 581–598.

Manzini, P., and M. Mariotti. 2014. "Welfare Economics and Bounded Rationality: The Case for Model-Based Approaches." *Journal of Economic Methodology* 21 (4): 343–360.

Marschak, J. 1960. "Binary-Choice Constraints and Random Utility Indicators." In *Stanford Symposium on Mathematical Methods in the Social Sciences,* edited by K. Arrow. Stanford University Press.

Matêjka, F., and A. McKay. 2015. "Rational Inattention to Discrete Choices: A New Foundation for the Multinomial Logit Model." *American Economic Review* 105 (1): 272–298.

McGeer, V. 2001. "Psycho-Practice, Psycho-Theory and the Contrastive Case of Autism: How Practices of Mind Become Second-Nature." *Journal of Consciousness Studies* 8 (5–6): 109–132.

McGeer, V. 2002. "Enculturating Folk Psychologists." *Synthese* 199 (1): 1039–1063.

McKelvey, R. D., and T. R. Palfrey. 1995. "Quantal Response Equilibria for Normal Form Games." *Games and Economic Behavior* 10 (1): 6–38.

Nogee, P., and F. Mosteller. 1951. "An Experimental Measure of Utility." *Journal of Political Economy* 59 (5): 371–404.

Polisson, M., and L. Renou. 2016. "Afriat's Theorem and Samuelson's 'Eternal Darkness'." *Journal of Mathematical Economics* 65:36–40.

Reny, P. J. 2015. "A Characterization of Rationalizable Consumer Behavior." *Econometrica* 83 (1): 175–192.

Ross, D. 1994. "Real Patterns and the Ontological Foundations of Microeconomics." *Economics & Philosophy* 11 (1): 113–136.

Ross, D. 2005. *Economic Theory and Cognitive Science: Microexplanation.* MIT Press.

Ross, D. 2009. "Ontic Structural Realism and Economics." *Philosophy of Science* 75 (5): 732–743.

Ross, D. 2012. "The Economic Agent: Not Human, But Important." In *Economics,* edited by U. Mäki, 691–735. Handbook of the Philosophy of Science, Volume 13. Elsevier.

Ross, D. 2014. *Philosophy of Economics.* Palgrave Macmillan.

Starmer, C., and R. Sugden. 1989. "Probability and Juxtaposition Effects: An Experimental Investigation of the Common Ratio Effect." *Journal of Risk and Uncertainty* 2 (2): 159–178.

Swait, J., and A. A. J. Marley. 2013. "Probabilistic Choice (Models) as a Result of Balancing Multiple Goals." *Journal of Mathematical Psychology* 57 (1–2): 1–14.

Tversky, A. 1969. "Intransitivity of Preferences." *Psychological Review* 76 (1): 31.

Wakker, P., I. Erev, and E. U. Weber. 1994. "Comonotonic Independence: The Critical Test Between Classical and Rank-Dependent Utility Theories." *Journal of Risk and Uncertainty* 9 (3): 195–230.

Wallin, A., J. Swait, and A.A.J. Marley. 2018. "Not Just Noise: A Goal Pursuit Interpretation of Stochastic Choice." *Decision* 5 (4): 253–271.

Wilcox, N. T. 2008. "Stochastic Models for Binary Discrete Choice Under Risk: A Critical Primer and Econometric Comparison." In *Risk Aversion in Experiments,* edited by J. C. Cox and G. W. Harrison. Research in Experimental Economics, volume 12. Emerald.

Wilcox, N. T. 2011. "Stochastically More Risk Averse: A Contextual Theory of Stochastic Discrete Choice Under Risk." *Journal of Econometrics* 162 (1): 89–104.

11 Real Patterns Analysis and Causality Neutrality

Harold Kincaid and Ryan M. Nefdt

11.1 Introduction

Daniel Dennett's body of work contains decades of countless insights into the behavioral sciences, the philosophy of science, and a number of adjacent arenas. His *intentional stance* (Dennett 1978, 1987) offers a genuinely novel interpretationist account of the mind in terms of how the adoption a particular kind of vocabulary toward agents and systems, both natural and artificial, can shed light on their behavior. However, despite the clear benefits of the position, ontological questions beset this kind of instrumentalism. Dennett (1991) had a subtle answer. What he proposes is a kind of "semi-realism" wedged loosely between hardcore realism and eliminative materialism, namely *real pattern* analysis.

The central question raised by this chapter concerns the practical significance of real patterns to social and behavioral science, where *causal* inferences are paramount. The idea of real patterns is innovative and suggestive, but it still is an open question as to how the idea applies in particular cases and what its relation to causal analysis is. Our basic strategy is this: First we attempt *faithful*[1] extensions of the idea to both modeling in behavioral science and formal representation in formal and computational linguistics, respectively. We argue that the fecundity of the analysis lies in its "causality neutrality"—that is, it is compatible with multiple causal stories for the given target phenomenon. Given the divergence in the nature of the case studies, we tentatively conclude that in spite of valiant attempts by others (Ladyman and Ross 2007; Anderson 2017; Kalis and Borsboom 2020) to link real patterns to causality, real patterns simply do not wear their causal structures on their sleeves. Furthermore, given the healthy naturalistic scepticism about causality in general, this might indeed be a virtue.

1. What counts as "faithful" is a matter of debate. Here we interpret it as a description as close to Dennett's (1991) information-theoretic characterization as possible.

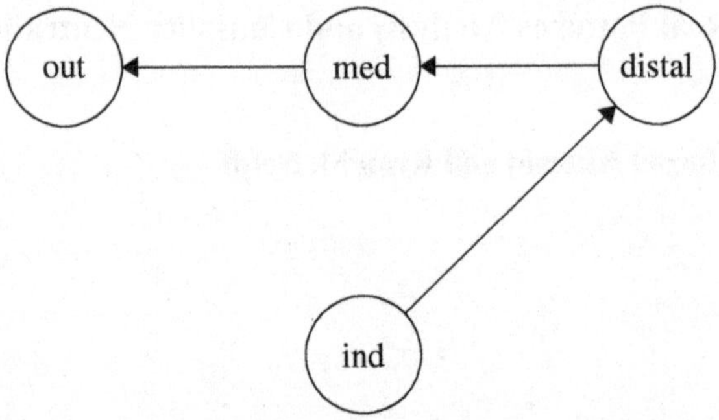

Figure 11.1: A causal structural model, where med is a mediating variable and direct cause of out, and ind is a direct cause of distal.

There is, of course, ongoing debate about what "real patterns analysis" comes to exactly (see the rest of the volume for more). As previously mentioned, the seminal idea comes from Dennett (1991). Firstly, a pattern is distinguished from a random string (or bitmap of data). Then, a pattern "is real—if there is a description of the data that is more efficient than the bitmap, whether or not anyone can concoct it" (Dennett 1991, 34). Ladyman and Ross (2007) extend these ideas by making it explicit that patterns need to be nonredundant to be real and that they must be projectable or generalizable. Real patterns are supposed to be exhibited above all by "fundamental physics" which is allegedly noncausal. Following Wilson (2006) and Morrison (2000), there is an interesting question whether there is "*a* fundamental physics." There are multiple, sometimes seemingly inconsistent models. And Frisch (2014) raises serious doubt about whether "basic" physics is noncausal. But we think these issues can be set aside if we agree that there are real patterns in the social and behavioral sciences, our target. Our question then is how do we apply and extend the notion of real patterns in these special sciences and what is their relation to causal inference. Section 11.1 looks at these issues as they surface in causal structural models[2] in the social sciences. Section 11.2 applies real patterns to formal linguistics or classical formal language theory. Our conclusion, in section 11.3, reaffirms the causal agnosticism or causality neutrality of Dennett's initial insight possibly suggesting a level of abstract real patterns distinct from the causal mechanisms associated with explanations in the special sciences.

2. "Structural" is used in multiple ways. Here it refers to describing what causes what but not to giving a parametric specification of the functions embodying the structure.

11.2 Real Patterns and Causality in the Social and Behavioral Sciences

This section looks at the relation between causal structural models that are widely used in the social and behavioral sciences and elsewhere and at their connection to the ideas of real patterns. There has been an explosion of work on the logic of causal structural models over the last twenty years (Pearl 2009). The progress has been on two fronts: top down and bottom up.

Top-down strategies show the implied probabilistic conditional dependencies and probabilistic independencies of a causal model, based on minimal assumptions about causality such as roughly that correlations that are not screened off represent causes and that there are no perfectly offsetting causes. With minimal assumptions, these can form the causal structure of a proposed causal model. Consider the model in figure 11.1. The model entails the following conditional probabilistic independency relations:

distal ⊥ out | med
ind ⊥ med | distal
ind ⊥ out | med
ind ⊥ out | distal

A test of the model is to see whether these implied probabilistic relations hold. In the practice of statistical analysis (usually involving multiple regression and maximum likelihood), care is needed in determining which variables should be conditioned on. Common causes should be conditioned on and "colliders" (where in a relation A→B←C the collider is B) should not. The logic of this is well worked out (e.g., Pearl 2009).

Not surprisingly, it is sometimes the case that there are other causal models that imply the same set of correlations, thus there is a case of underdetermination in philosophy of science lingo. Thus, the model of figure 11.1 is consistent with a set of other models as illustrated by figure 11.2, where the links without arrowheads suggest that the causal directions are indeterminate—the probabilistic relations are not enough to decide on causality.

The bottom-up approach to causal inference starts with the data without assuming a causal model and tries to find a model consistent with the data. It is not without assumptions, but they are the basic ones relating causes and probabilities used in top-down approaches—for example, causes imply conditional dependencies. Then algorithms are developed to systematically search variable by variable for probabilistic dependencies and independencies that are signs of causation. There are many such algorithms. Tested against simulated data where the true causal relations are known, these algorithms have a high success rate.

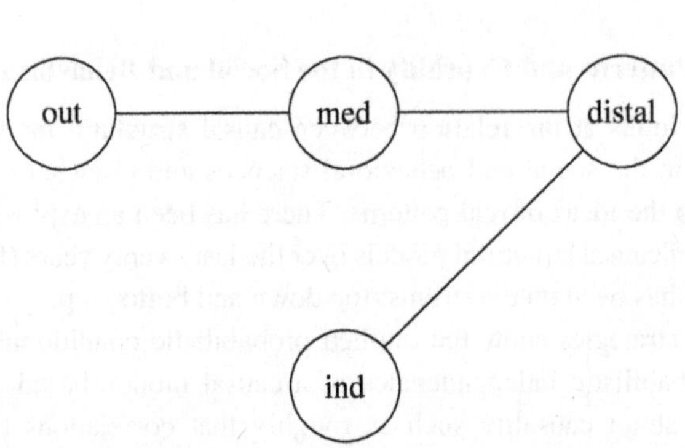

Figure 11.2: The equivalence class of the causal model in figure 11.1.

With this background, we can turn now to ask what is the connection of causal structure to real patterns. There are two basic questions: (1) If we have well-confirmed causal structural models, what do they entail about real patterns? and (2) What do well-confirmed real patterns tell us about structural causal relations?

Obviously, a successful causal model—whether by deductive or inductive criteria—entails regularities in the form of probabilistic relations between variables. In a good causal model the regularities are nonredundant—they cannot be explained by other variables such as common causes—and no other simpler model of the causal relations explains the data.

However, as we saw above in figure 11.2, it is not unusual for there to be multiple causal models consistent with the data. But if we have good evidence for a causal model, we have good evidence for the real patterns that constitutes its equivalence class—the patterns illustrated in figure 11.2. Furthermore, it is also possible to determine what causal relations hold across all models consistent with the data. These causal relations entail real patterns. An obvious proposal is that areas of scientific endeavor where finding causes is paramount, a stronger notion of real patterns is that of real causal patterns. Of course, some good science is satisfied when it finds stable associations—correlations that are generalizable and nonredundant. The use of differential equations in applied physics is an example. However, we think there is often more to science than just finding stable associations. Finding a causal mechanism is a good route to confidence that your correlations will hold up in new circumstances.

Turning to question (2)—what do well-confirmed real patterns tell us about structural causal relations—we see further appeal to talking about real causal patterns. The reason is fairly simple: Looking only for real patterns leads not only to missing real causal relations, but more worryingly, it leads to systematic causal error. At an abstract level this worry can be motivated by intuitive slogans such as "No causes in, no causes out" (Cartwright 1989) and the "Correlation is not causation" mantra found in indefinitely many statistics texts. The revolution in causal inference shows that these claims are too simplistic. Correlations can show causation in the right circumstances; the causal assumptions in the case of bottom-up discovery are minimal. However, the slogans are on the right track.

Nonetheless, such generalities are not as compelling as concrete demonstrations. So here is one that gives a specific sense to the abstract notion of real patterns in practice and shows that searching for real patterns makes systematic mistakes in identifying real causal patterns.

The Aiake Information Criterion is used across the biological and social and behavioral sciences (Burnham and Anderson 2002). It is built on the Kullback–Leibler (KL) divergence measure of information. What it adds in addition is a simplicity term—basically models with fewer variables are scored higher. The formula is:

$$\mathrm{AIC} = 2\mathrm{K} - 2\ln(\mathrm{L}), \tag{11.1}$$

where L is the likelihood of the data and K is the number of variables. This criterion is usually used to evaluate models estimated by maximum likelihood, which cycles through various parameter values until a best fit model is found—one that best predicts the data. The penalty for the number of variables is sometimes thought to be arbitrary. However, in this context it can be seen as a concrete attempt to eliminate redundant patterns due to overfitting; in addition, AIC performs well in tests against simulated data with known properties.

AIC thus uses informational value and searches for the simplest model as well. It is one way to make concrete the idea of finding a non-redundant real pattern. The question then in the context of this discussion is how the AIC version of real patterns relates to causality.

To ask that question, data were simulated with a known causal structure. The structure used is designed to illustrate a full range of possible causal relations: direct causes that work through no other variable, mediators that lie between a distal cause and an effect, confounders (causes of two independent variables), and colliders (variables which are caused by two independent variables). In statistical practice conditioning on common causes eliminates spurious correlations, conditioning on mediators removes true correlations, and conditioning on colliders creates spurious correlations. The effect sizes of all the relations were specified in generating the data. A dataset of one hundred thousand

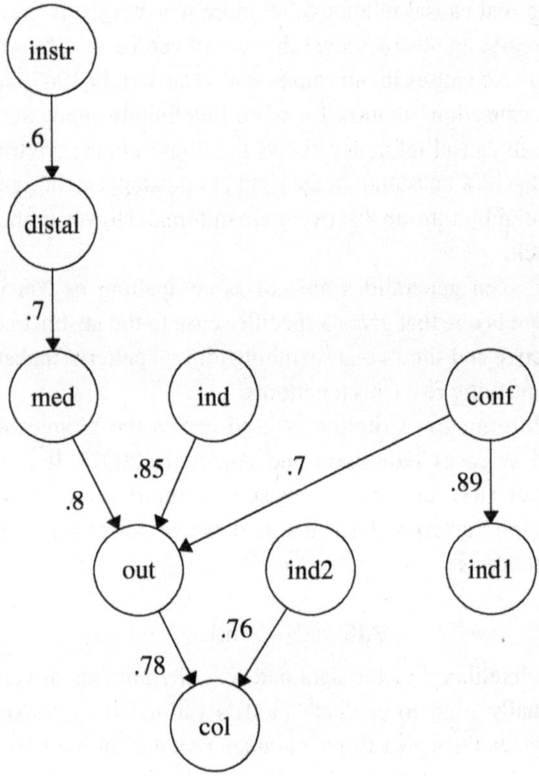

Figure 11.3: The causal model used to simulate the data.

observations was generated with standard nice statistical properties (e.g., i.i.d. errors). Figure 11.3 illustrates the causal structure used in the simulation.

Once the data are simulated, the next step is to see what the AIC criterion tells us about which variables should be included in a model of the simulated data. The question is whether an informational criterion as embodied in AIC pick out the causes. The variables that the AIC algorithm[3] selected are in table 11.1 as well the effect sizes (the betas). There are three things to note:

1. Distal and instr are dropped as causally relevant when they are.
2. Ind2 and col are cited as significantly causally relevant when they are not.
3. The correct causal inclusions have noticeably smaller effect sizes than is the case in the data.

3. We use the OLSRR R package of Hebbali (2020), which is a form of stepwise forward regression.

Table 11.1
Results of applying an AIC algorithm to the simulated data where the algorithm is telling us which variables ought to be included and their relative importance (*Beta*). *Sig* is the significance level of the postulated relationships. Lower and upper confidence intervals are represented by *lower* and *upper*.

Model	Beta	SE	Std. Beta	t	Sig	Lower	Upper
col	0.485	0.002	0.500	247.246	0.000	0.481	0.489
ind2	-0.369	0.002	-0.274	-189.999	0.000	-0.373	-0.366
ind	0.529	0.002	0.394	294.566	0.000	0.525	0.532
conf	0.434	0.002	0.321	263.725	0.000	0.431	0.438
med	0.498	0.002	0.274	239.170	0.000	0.494	0.502

So following this version of real patterns produces systematic errors when it comes to picking out causes.[4]

It may well be that this AIC algorithm is not the best way of picking out real patterns. But that would then still leave the question of what methods are better? How do we turn the intuitively appealing idea of real patterns into real-world applications? And, what would a better method say about causation? Even if there are better methods, the AIC case has still established a conceptual distinction between causal models and real pattern analysis in statistical social science.

We turn next to look at another case of applying real patterns while remaining agnostic about the specific causal implications of the theory.

11.3 Dennett Meets (Early) Chomsky

In the previous sections we largely held the causal structural models fixed while asking probing questions about their status as real patterns. In the following sections, we attempt the reverse. Specifically we will define a class of formal models with explicit real patternhood and ask what causal mechanisms (if any) it entails. This section comes with a serious caveat. We are NOT saying that early formal linguistics is or was the correct way to identify real patterns in language (or even the only way). Rather we are saying that the tools from formal language theory are the closest translations of the information-theoretic ideas of compression, redundancy, and pattern representation present in Dennett's (1991) vision of real patterns. We use these tools to establish conceptual

4. Our doubts about the use of AIC here is not an argument about its use everywhere. If explicitly structural causal models are being tested, then AIC may well be a fruitful criterion for model selection. See Shipley (2013).

distance from the predictive and explanatory power of real pattern analysis and causal theories of the production and comprehension of language. Whatever further theory you endorse should not contradict the findings of FLT, which form an underlying structural picture of syntax.

The first task is to harness the initial insights of Dennett (1991), in terms of information theory and compressibility, into an analysis of a realm of patterns surprisingly left out of debates on the subject, namely formal linguistics and natural language. In section 11.3.1, we briefly describe Chomsky's initial insights into formal language theory (FLT), the notion of a formal language, grammar, and accompanying automata. Thereafter, we translate some features of Dennett's (and others', such as Ladyman and Ross's [2007]) information-theoretic characterization of real patterns into this discrete system of describing linguistic structure. Despite the congruence with real pattern analysis, we argue that FLT lives at the level of representation, generates predictions, and might even constitute a certain kind of "linguistic stance" while remaining noncommittal about the causal mechanisms at the heart of natural language.

11.3.1 Grammars and Patterns

The core innovation that emanated from early mathematical linguistics was the application of proof theory to natural language syntax. Chomsky, leaning into the work of Turing, Church, and Post among others, produced a concept of a grammar as a string rewriting system (Chomsky 1957, 1965). The grammar, made up of a finite set of rules for such conversions, involved capturing or compressing patterns of complex linguistic forms into formats familiar from the kinds of sequent calculi of natural language deduction (especially of the Gentzenian variety).

More formally, a grammar G consists of four core elements: (1) a set of terminal vocabulary V_T (which you can think of as words), (2) a set of nonterminals (or syntactic categories) V_N, (3) a set of production rules P, and (4) a unique start symbol S, or $G = (V_T, V_N, P, S)$. The total vocabulary of G is given by the disjoint union of V_T and V_N with V^* denoting the set of all possible strings of vocabulary elements (Levelt 2008). The production rules are ordered pairs of strings (from V) or the Cartesian product of V^+ (a set that excludes the null string) and V^* ($V^+ \times V^*$). This means that P licenses the replacement of strings from V^+ with strings from V^*, or rather strings from V^+ can be *rewritten* as strings from V^*, hence the rewriting system. S is a subset or specific element of V_N. The game is then to derive the strings of natural language from the combinatorics of the alphabet and a finite set of rules, or the grammar. Jäger and Rogers (2012, 1957) describe the picture in the following way:

> *G* will be said to *generate* a string *w* consisting of symbols from if and only if it is possible to start with *S* and produce *w* through some finite sequence of rule applications. The sequence of modified strings that proceeds from *S* to *w* is called a *derivation* of *w*. The set of all strings that *G* can generate is called the language of *G*, and is notated *L*(*G*).

Different kinds of grammars can generate different sets of strings or rather stringsets of different complexity. We'll return to this feature below. Chomsky's (1957) definition of a $[\Sigma, F]$ grammar is a particular, Post-inspired (see Pullum 2011), version of this procedure:

$[\Sigma, F]$ Grammar: Each such grammar is defined by a finite set Σ of initial strings and a finite set F of "instruction formulas" of the form $X \rightarrow Y$ interpreted: "Rewrite *X* as *Y*." (22)

These "instruction formulas" or production rules can be thought of as essentially compressions of the data (see Nefdt 2023). They identify syntactic regularities such as the fact that some English sentences consist of a noun phrase and a verb phrase as in "Dan wrote the book" or $S \rightarrow NP + VP$. In fact, in the erstwhile Standard Theory of generative grammar, a procedure involving such "phrase structure rules" (coupled with certain transformation rules for converting constructions like passives) was thought to formally capture the syntax of natural language. This formalism even covers tricky kinds of syntactic patterns. For instance, natural language syntax seems to allow for multiple embedding and subordinating structures such as the following:

1. A man that a woman loves just walked in.
2. Dan believed that Davidson said that Churchland held that the mind is an illusion.
3. Susan is very very tired.

Linguistic structures iterate and combine to form more complex structures, seemingly ad infinitum. This property requires rules that can capture embedding and subordination in the syntax. Hence, phrase structure grammars were often associated with context-free grammars (and sometimes more complex context-sensitive ones, see Manaster and Kac 1990). A context-free grammar has the following set of rules, where a, b $\in V_T$:

a: S $\rightarrow$ ab
b: S $\rightarrow$ aSb

Thus, sentences or strings can be constructed from other strings, allowing for endless iteration or recursion. Drawing from early computability theory, and assuming the set of natural language sentences is potentially infinite, recursion became an essential part of the formal apparatus of grammar (Tomalin 2006).[5]

Nevertheless, two issues plagued early formal linguistics: (1) the issue of compressibility, and (2) the related problem of complexity. In terms of compressibility, identifying the patterns of natural language via this process of phrase structure rewriting rules was largely inefficient. Better compression algorithms were needed. As Bickerton (2014, 24) states, "Rule proliferation and 'ordering paradoxes' were only two of a number of problems that led to the eventual replacement of the Standard Theory." In other words, there were just too many phrase structure rules (for each construction of each language) and what side of the rewriting arrow a string belonged to could be justified in either direction. Seen in this light, the Extended Standard Theory (Jackendoff 1977) was a triumph of compressibility. All the (hundreds) of phrase structure rules could simply be characterized by or reduced to three types of schemata known as X-bar theory collectively, which looked like: $X' \rightarrow X(WP)$ or $X' \rightarrow (WP)X$ (i.e., the complement rule) where X, WP are phrasal categories. More contemporary generative theories such as Minimalism (Chomsky 1995) aim to reduce all of syntax (and by definition the narrow language faculty itself) to the further simplified *Merge* operation, which takes two syntactic objects and produces a labeled set containing both as members.[6] It seems that one way to characterize linguistic theory (or a certain variety thereof) is as an ongoing struggle to more efficiently compress the rules of language (for various reasons such as learnability and evolutionary plausibility inter alia).

Interestingly, compression of this nature directly confronts complexity in FLT. The initial game of formal linguistics was to capture "all and only" the strings of natural language or, in Dennettian terms, the real patterns of language. But in attempting such compression, the formalisms suffered from overgeneration problems. Both X-bar and Merge generate more structures or patterns than are linguistically realized in known human languages. In fact, complexity concerns were there from the start. As we saw above, phrase structure grammar was assumed to be context-free. This is partly because context-free languages are rather well behaved in terms of complexity and parsability (but still capture embedding phenomena, see Chomsky 1956). They sit above the regular languages in the original Chomsky Hierarchy (see figure

5. And later made its way into the very ontology of language, see Hauser et al. (2002), Berwick and Chomsky (2016), Chomsky et al. (2023).

6. If you are worried about where recursion went, Merge can be applied iteratively to its own output thereby retaining the property.

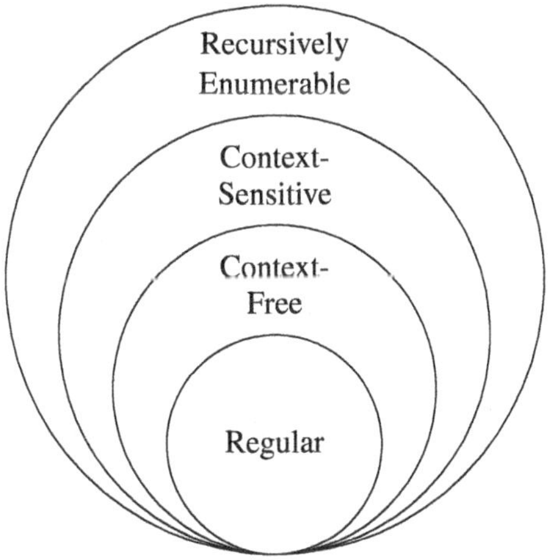

Figure 11.4: The Chomsky Hierarchy.

11.4), where each ring above this class of formal languages increases the complexity of the patterns produced (and subsumes in one below). Peters and Ritchie (1973) showed that the transformational grammar of the Standard Theory generates the set of recursively enumerable languages: It is a Type-0 or unrestricted grammar with undecidable complexity.

In practice, X-bar and Minimalism adopted a number of linguistically motivated constraints so as to produce or constrain only the real patterns of language (and nothing more). In terms of complexity, due to a number of proofs involving cross-serial dependency structures (e.g., *John the man drowned saw*) in languages such as Dutch and Swiss-German, an entirely new ring of the Hierarchy (between context-sensitive and context-free) had to be established in which to locate natural language syntax, namely the "mildly context-sensitive languages" (Shieber 1985; Joshi et al. 1991). The real patterns of natural language were mildly context-sensitive, the grammars recursive, and the structures parsable by human minds (in polynomial time).

On a separate but related note, Dennett (1991) (and elsewhere) emphasizes the significance of the fascinating mathematical configurations of Conway's *Game of Life* (GoL) for an understanding of real patterns. He uses this famous cellular automaton as an example of a kind of perspectival realism or "intellectual" patterning. Ladyman and Ross's (2007) "scale relativity of ontology" takes much inspiration from the idea that certain *real* patterns emerge at different levels of systematic description, such as those found in the special sciences.

At some base level of description, the GoL is nothing but a discrete binary checkers board on a 2D grid. The dynamics of the system are given by simple collocational rules (*life* and *death* determined by the populations of neighboring cells). Yet at other (perhaps higher) levels, there appears to be movement, dynamism, and an apparent facsimile of life. Scientists and hobbyists alike have spent decades identifying such patterns. Further, the GoL has been shown to reveal features characteristic of an "omniperiodic" system—that is, there is a pattern that repeats itself (called an "oscillator") after n steps for every positive integer, where $n \in Z^+$ (Brown et al. 2023).

What's interesting from our point of view, and to further drive the real pattern analysis home, is the correspondence between classes of formal languages and distinct automata. Each ring of the Chomsky Hierarchy can be associated with a special accepting machine or automaton. The class of least complex languages, regular languages, are associated with standard finite state machines or automata. Context-free languages match up with pushdown automata, which incorporate a "memory" stack for accepting strings like $a^n b^n c^n d^n$. This is part of the reason that FLT has been applied well outside of the bounds of the language sciences. States need not be words, and transitions need not be linguistic categories. They could be programs, or cells, or neurons, agents, and so on. What might seem static from the perspective of formal grammar can be represented dynamically by means of the accompanying automata accepting strings of the generated languages.

There is a lot more one can say about FLT, automata, the nature of the Chomsky Hierarchy, computational complexity, and so on. Much of these details are beyond the present scope (see Levelt [2008] for an accessible guide). What is important from the above discussion is that the study of formal grammar in linguistics can readily be construed as an exercise in compression and representation akin in nontrivial ways to the kinds of claims and arguments used to motivate real pattern analysis by Dennett and others.

Despite the connections, one salient feature of FLT that separates it from the kinds of discussions of real patterns found in the literature is the fact that FLT trades in discrete as opposed to continuous mathematical representation. The kind of information-theoretic insight that Dennett brought into both the philosophy of mind and the philosophy of science crucially relies on results in coding theory such as Kolmogorov (1963), which links the compressibility of data with the existence of statistical regularities within those data.[7] Kolmogorov complexity itself, which defines the complexity of a finite string by

7. In fact, in the other direction, formal linguistics has been rather dismissive of the significance of statistical tools and models in the study of language, see Pullum (2007).

the length of the shortest program for a reference Universal Turing Machine (UTM) that outputs that string, has become a useful metric in the philosophy associated with real patterns (see Suñes and Martínez 2021). But talk of bits, noise, pixels, and compression introduces statistical representations that seem, at least initially, to be alien to the discrete representations of FLT. This appearance is, of course, only perfunctory. As Piantadosi (2023) remarks, "The key is that discreteness is a special case of continuous modeling, meaning that theories which work with continuous representations get the best of both worlds, fitting discrete patterns when appropriate and gradient ones otherwise" (18).

Take Dennett's (1991) initial concept of compressibility. He states that "A series (of dots or numbers or whatever) is random if and only if the information required to describe (transmit) the series accurately is incompressible" (32). This is the essence of formal linguistics—that is, the assumption that a series of strings is not random with relation to syntactic structure. Compare Hunter's recent characterization of the study of formal grammar or FLT:

> This new branch of mathematics provided the formal grounding for a new conception of linguistics in which grammars, rather than sentences or collections of sentences ... The observed regularities come to be seen as consequences of the structure of the underlying system, the grammar. (2021, 74)

In other words, grammars compress linguistic regularities and produce a method for generating the compressed set (the definition of a compression algorithm [Millhouse 2021]). In fact, most (generative) linguists assume that the syntax of natural language constitutes a countably infinite set (or array or collection) only capable of decompression via the proof-theoretic means discussed above (or some version of it). As Chomsky et al. (2023) put it, "A driving goal of the modern study of language is to determine and explain this property of discrete infinity" (1).[8]

Similarly, Ladyman and Ross (2007) introduce the notion of *indispensability* to real patternhood in order to add clarity where Dennett's model might be ambiguous. According to them, a theory of real patterns should help decide which patterns (partial encodings of the dataset) are the real ones from sets of mutually redundant options. Their idea is that the indispensable patterns are the real ones. In FLT, there are similar issues. A number of distinct grammars generate the same stringsets. They are "weakly equivalent" in the parlance of FLT. Quine (1972) even used this fact to challenge the mentalism of generative

8. See Pullum and Scholz (2010) and Nefdt (2019) for some issues with the linguistic infinity postulate.

grammar. In other words, if various different grammar formalisms all generate the same sets of sentences, then how can we determine which of the many possible grammars is the one cognized by the language user. Underdetermination of theory by data under a different guise, it seems. However, despite the threat posed by these weak equivalence proofs to the real patterns potentially unearthed by formal grammars, indispensability can still be salvaged in FLT. The strategy taken in Nefdt (2023) is to look to the notion of strong generative capacity. Different grammar formalisms such as phrase structure, dependency grammar, and categorial grammar might generate the same stringsets (have the same "weak generative capacity") but they fail to pick out the same structures.

The strong generative capacity of a grammar is given by the kinds of mathematical objects it designates. For example, phrase structure grammars are generally represented by hierarchical tree structures with constituents, while dependency grammars are represented by flat, rooted acyclic graphs without an obvious concept of constituency (they are so-called lexicalized grammars—i.e., they only have VT and no VN). These are the "structural descriptions" linguists often talk about. Chomsky (1965) even states that "discussion of weak generative capacity marks only a very early and primitive stage of the study of generative grammar. Questions of real linguistic interest arise only when strong generative capacity (descriptive adequacy) and, more important, explanatory adequacy become the focus of discussion" (61). Thus, there might be loss (or a gain) of structural information when moving from one formalism to the other even if the same sets of strings are generated. One formalism encodes constituency (phrase structure) while the other formalism encodes semantic argument structure (dependency). Both generate the same sets of strings nonetheless. Or in other words, the strong generative capacity of formal grammars provides potentially indispensable information in accordance with Ladyman and Ross's account of real patterns.[9]

Again, the fuller details are presented in Nefdt (2023), here we are only aiming for a lossy illustration. The basic point is that formal language theory offers a direct proxy for real pattern analysis in terms of discrete representations of natural language.[10]

9. Millhouse's (2021) account of model similarity as a metric for compressibility can also be represented in FLT in terms of finite state transducers, which directly compare grammars and their automata.

10. An alternative route might be to follow the line of Chater et al. (2015) in converting discrete linguistic structures into probabilistic ones through the model selection tool of minimum description length.

11.3.2 Causal Structures in Language

Problems arise when we investigate the causal mechanisms of natural language, either in the human mind/brain or the linguistic behavior of language users in communities. We will briefly highlight two such issues, one from early attempts at finding a psycholinguistic grounding for formal linguistic structures and the other from a quixotic philosophical argument presented in Davidson (1986).

Psycholinguistics as a disciplinary enterprise emerged from the ideal of finding formal linguistic structure at the level of psychological mechanisms. For example, one hypothesis of early psycholinguistics was the "derivational theory of complexity" or the claim that "the complexity of a sentence is measured by the number of grammatical rules employed in its derivation" (Fodor et al. 1974, 319). In spite of some initial excitement with relation to click location experiments (designed to identify constituent boundaries in sentences [Holmes and Forster 1972]) the eventual conclusion was that no place for internalized grammar rules could be found in actual parsing: "There exists no suggestions about how a generative grammar might be concretely employed as a sentence recognizer in a psychologically plausible model" (Fodor et al. 1974, 75). The causal structures of language seemed independent of or at least recalcitrant to the real patterns described at the formal level. Eventually, psycholinguistics established itself as a discipline largely independent of theoretical linguistics. A similar story can be told for neurolinguistics. As Baggio (2022) notes, compared to phonology, "the attempt to connect formal linguistic operations to specific brain processes has been less successful" (59) and when it comes to neurolinguistics "concrete modeling tends to take precedence over abstract theory" (63). Even much publicized results such as Ding et al. (2016), which purport to provide MEG data indicating that cortical responses at different timescales, track constituents at different hierarchical levels: that is, words, phrases, and sentences, are challenged by the empirical equivalence of using frameworks involving only neural networks with sequentially represented vectors, in other words, no hierarchical constituent structure (Frank and Yang 2018).

It seems that psycholinguistics and neurolinguistics, the scientific studies of causal algorithms and mechanisms at the heart of language comprehension and production, respectively, either say little about abstract real patterns or are compatible with multiple such formalisms. The puzzle is that FLT (and other kinds of formal theories) provide useful predictions (e.g., expect a VP to follow an NP in English). It can explain both the complexity and compressibility of natural language syntax to a significant degree. And yet it is unclear what causal mechanisms these representations map on to or even suggest at the best of times. Chomskyans have often adopted a wait-and-see approach to eventual unification with cognitive neuroscience. Perhaps this is cautiously optimistic. But the trends in contemporary neuroscience (or natural language processing for that matter) do not seem to be converging on the incorporation of formal linguistic structure but rather further and further away from it.[11]

On the more philosophical side, Davidson's (1986) deconstruction of formal models of language is illustrative of the general concern. Let's start with his remarkable conclusion:

> I conclude that there is no such thing as a language, not if a language is anything like what many philosophers and linguists have supposed. There is therefore no such thing to be learned, mastered, or born with. We must give up the idea of a clearly defined shared structure which language-users acquire and then apply to cases. (Davidson 1986, 446)

One interpretation of his journey to the above position is as a point about real patterns. The idea is something like this: There are two different kinds of theory (or stance) a given hearer could use to interpret a given speaker's utterances, a "prior theory" and a "passing theory." The prior theory is of the systematic nature of FLT (or Davidson's own work on semantic theory). But when we actually encounter each other in conversation, we don't seem to use such patterns for prediction at all. This claim curiously dovetails with Fodor et al.'s psycholinguistic conclusion above. What we do use, according to Davidson, is a series of momentary heuristics or a passing theory, a folk linguistics if you will. The two theories do not seem compatible. A speaker of English might flout all the carefully crafted grammar rules of the frustrated linguist and yet still communicate efficiently. In fact, following Stainton (2016), the prior theory becomes explanatorily otiose if it is unutilized with relation to language use. Put in another way, the real patterns described by the formal theory do not figure into the explanation or practice of natural language as it is found in linguistic communities of speakers.

There seem to be at least two strategies open to Dennettians here. The first involves identifying distinct layers of representational stances. This corresponds to the relationship between the intentional, design, and physical stances. Davidson's passing theory is at the level of the intentional stance. FLT most likely lives at the physical level. Causation is perhaps at the level of the design stance. The problem is that psycholinguistics, neurolinguistics, and the passing theory are all often considered causal models in the special sciences. How they link up to the prior theory (where the real patterns seem to live) is a

11. Perhaps with the exception of emerging fields like BERTology in NLP (Rogers et al. 2021).

matter open for investigation. This shouldn't be too surprising if we take Dennett's *Game of Life* analogy seriously. There is no movement ("amoeba-like" as Dennett calls it), collision, consumption, or cause at the level of cellular automata. Shifting ontological perspectives introduces different levels of description.

> Note that there has been a distinct ontological shift as we move between levels; whereas at the physical level there is no motion, and the only individuals, cells, are defined by their fixed spatial location, at this design level we have the motion of persisting objects; it is one and the same glider that has moved southeast in figure 11.2, changing shape as it moves, and there is one less glider in the world after the eater has eaten it in figure. (Dennett 1991, 39)

The second available option is to flesh the theory of causation (or design level) out more. Jackendoff's (2002) parallel architecture attempts to do exactly this. It jettisons the centrality of structural or syntactic analysis characteristic of early generative theory, and offers a more comprehensive interconnected generative picture with the explicit aim of bridging the gap between linguistic theory and psycho/neurolinguistics. But importantly, it does not contradict the proofs of FLT on the patterns and complexity of natural language syntax. In addition, adding more causal mechanisms, phonological, semantic, and pragmatic information can move us further from the compressed representation of Dennett toward a more full "bitmap" (i.e., larger file size) and thus further from the minimalist real pattern analysis offered by FLT.[12]

Of course, a stronger eliminativist position on causation has been urged by many naturalistic philosophers. On these views, fundamental physics does not seem to trade in causes. Our intuitions might, but so much the worse for our intuitions. Interestingly, Ladyman and Ross (2007) adopt a different strategy despite their adherence to heavy-duty naturalism. They distinguish between the folk concept, the philosophical concept, and the notional-world concept of causation. Only the last of these is relevant for science, they argue. Briefly, "A notional world is a network of background assumptions that permits, but also restricts, projections of objects by reference to their utility to some set of practical purposes" (Ladyman and Ross 2007, 244). In Dennett (2017), he calls this idea an "umwelt" or restricted ontology required by a given organism or machine for its operations (and not much more). Ladyman and Ross caution against naturalistically inspired eliminativism about causes (even though

12. Something like Jackendoff's account might indeed be correct for natural language. Nothing we say relies on the truth of Chomsky's early theory (which he himself mostly rejects now). In a sense, FLT is a discrete mathematical framework for achieving compression of syntactic structure. Its continued use in computational linguistics is testament to that fact.

they are sympathetic to the position). For instance, they argue that the special sciences successfully track real patterns by "book-keeping them as individuals interacting in causal processes" (290). This conclusion is generally compatible with our concept of causal neutrality. However, it does not rule out the possibility of a level of analysis independent of causal processes or mechanisms in the special sciences, a formal level of structural representation aligned with the minimalist information-theoretic machinery of Dennett's (1991) picture.

As with the AIC model of the previous case study, FLT is not the only way to formalize linguistic or syntactic structure, but it is remarkably similar in nature to real pattern analysis offered by Dennett (1991), as we hope to have shown. Furthermore, causality neutrality seems to be apparent in both cases.

11.4 Conclusion

In this chapter, we have extended real pattern analysis into two distinct domains, causal modeling in social science and formal linguistics in cognitive science. In the first case, we presented a picture in which the causal mechanisms are known in advance, then applied a particular kind of real pattern analysis to them. Our conclusion there was that the real *causal* patterns are distinct from real patterns *simpliciter* in some statistical social scientific models. The next case started with explicit or "literal" real patterns modeled on Dennett's (1991) information-theoretic account, in terms of early linguistic grammars. Here, we claimed that a hierarchy of stances is in fact possible with the most basic structural level exhibiting the property of causal neutrality.

Of course, our methodology has its limitations. We considered only two cases studies, albeit very different ones. Thus, our conclusions are more preliminary and illustrative of a call for further robust work into these and other domains where real pattern analysis can clearly make a significant philosophical difference and causality seems to play an essential role in that process.

References

Andersen, H. 2017. "Patterns, Information, Causation." *Journal of Philosophy* 114 (11): 592–622.

Berwick, R., and N. Chomsky. 2016. *Why Only Us? Language and Evolution.* MIT Press.

Bickerton, D. 2014. *More Than Nature Needs: Language, Mind, and Evolution.* Harvard University Press.

Brown, N., C. Cheng, T. Jacobi, M. Karpovich, M. Merzenich, D. Raucci, and M. Riley. 2023. "Conway's Game of Life Is Omniperiodic." *arXiv preprint,* eprint: 2312.02799.

Burnham, K., and D. Anderson. 2002. *Model Selection and Multimodel Inference.* Springer.

Chater, N., A. Clark, J. Goldsmith, and A. Perfors. 2015. *Empiricism and Language Learnability.* Oxford University Press UK.

Chomsky, N. 1956. "Three Models for the Description of Language." *IRE Transactions on Information Theory* IT-2:113–123.

Chomsky, N. 1957. *Syntactic Structures.* Mouton Press.

Chomsky, N. 1965. *Aspects of the Theory of Syntax.* MIT Press.

Chomsky, N. 1995. *The Minimalist Program.* MIT Press.

Chomsky, N., D. Seely, R. Berwick, S. Fong, M. Huybregts, H. Kitahara, A. McInnerney, and Y. Sugimoto. 2023. *Merge and the Strong Minimalist Thesis.* Cambridge University Press.

Davidson, D. 1986. "A Nice Derangement of Epitaphs." In *Truth and Interpretation,* edited by D. Davidson, 251–265. Blackwell.

Dennett, D. C. 1978. *Brainstorms.* Bradford.

Dennett, D. C. 1989. *The Intentional Stance.* MIT Press.

Dennett, D. C. 1991. "Real Patterns." *Journal of Philosophy* 88 (1): 27–51.

Dennett, D. C. 2017. *From Bach to Bacteria and Back.* Penguin Books.

Ding, N., L. Melloni, H. Zhang, X. Tian, and D. Poeppel. 2016. "Cortical Tracking of Hierarchical Linguistic Structures in Connected Speech." *Nature Neuroscience* 19 (1): 158.

Fisch, M. 2014. *Causal Reasoning in Physics.* Cambridge University Press.

Fodor, J., T. Bever, and M. Garrett. 1974. *The Psychology of Language.* McGraw-Hill Publishers.

Frank, S., and J. Yang. 2018. "Lexical Representation Explains Cortical Entrainment During Speech Comprehension." *PloS ONE* 13 (5).

Hauser, M., N. Chomsky, and W. T. Fitch. 2002. "The Faculty of Language: What Is It, Who Has It, and How Did It Evolve?" *Science* 298:1569–1579.

Holmes, V., and K. Forster. 1972. "Click Location and Syntactic Structure." *Perception & Psychophysics* 12:9–15.

Hubbell, A. 2020. *Olsrr: Tools for Building OLS Regression Models (0.5.3) [R].* Https://CRAN.R project.org/package=olsrr.

Hunter, T. 2021. "The Chomsky Hierarchy." In *Blackwell Companion to Chomsky,* edited by N. Allott, T. Lohndal, and G. Rey, 74–95. Wiley-Blackwell.

Jackendoff, R. 1977. *X Syntax: A Study of Phrase Structure.* MIT Press.

Jackendoff, R. 2002. *Foundations of Language.* Oxford University Press.

Jäger, G., and J. Rogers. 2012. "Formal Language Theory: Refining the Chomsky Hierarchy." *Philosophical Transactions of the Royal Society B: Biological Sciences* 367 (1598): 1956–1970.

Joshi, A., K. Vijay-Shanker, and D. Weir. 1991. "The Convergence of Mildly Context Sensitive Grammatical Formalisms." In *Foundational Issues in Natural Language Processing,* edited by P. Sells, S. Shieber, and T. Wasow, 31–81. MIT Press.

Kalis, A., and D. Borsboom. 2020. "Folk Psychology as a Causal Language." *Theory & Psychology* 30 (5): 723–728.

Kolmogorov, A. 1963. "On Tables of Random Numbers." *Sankhyā: The Indian Journal of Statistics, Series A (1961-2002)* 25 (4): 369–376.

Ladyman, J., and D. Ross. 2008. *Everything Must Go.* Oxford University Press.

Levelt, W. 2008. *An Introduction to the Theory of Formal Languages and Automata.* John Benjamins.

Manaster-Ramer, A., and M. Kac. 1990. "The Concept of Phrase Structure." *Linguistics & Philosophy* 13 (3): 325–362.

Millhouse, T. 2021. "Compressibility and the Reality of Patterns." *Philosophy of Science* 88 (1): 22–43.

Morrison, M. 2000. *Unifying Scientific Theories.* Oxford University Press.

Nefdt, R. 2019. "Infinity and the Foundations of Linguistics." *Synthese* 196 (5): 1671–1711.

Nefdt, R. 2023. *Language, Science, and Structure: A Journey into the Philosophy of Linguistics.* Oxford University Press.

Pearl, J. 2009. *Causality.* 2nd ed. Cambridge University Press.

Peters, S., and R. Ritchie. 1973. "On the Generative Power of Transformational Grammars." *Information Sciences* 6:49–83.

Piantadosi, S. 2023. *Modern Language Models Refute Chomsky's Approach to Language.* Lingbuzz Preprint. url: https://lingbuzz.net/lingbuzz/007180.

Pullum, G. 2007. "Ungrammaticality, Rarity, and Corpus Use." *Corpus Linguistics and Linguistic Theory* 3:33–47.

Pullum, G. 2011. "The Mathematical Foundations of Syntactic Structures." *Journal of Logic, Language & Information* 20 (3): 277–296.

Pullum, G., and B. Scholz. 2010. "Recursion and the Infinitude Claim." In *Recursion in Human Language,* edited by H. van der Hulst, 113–138. Studies in Generative Grammar 104. Mouton de Gruyter.

Quine, W.V.O. 1972. "Methodological Reflections on Current Linguistic Theory." In *Semantics of Natural Language,* edited by D. Davidson and G. Harman, 442–454. D. Reidel.

Rogers, A., O. Kovaleva, and A. Rumshisky. 2021. "A Primer in BERTology: What We Know About How BERT Works." *Transactions of the Association for Computational Linguistics* 8:842–866.

Shieber, S. 1985. "Evidence Against the Context-Freeness of Natural Language." *Linguist and Philosophy* 8:333–343.

Shipley, B. 2013. "The AIC Model Selection Method Applied to Path Analytic Models Compared Using a D-Separation Test." *Ecology* 94 (3): 560–564.

Stainton, R. 2016. "A Deranged Argument Against Public Languages." *Inquiry* 59 (1): 6–32.

Suñé, A., and M. Martínez. 2021. "Real Patterns and Indispensability." *Synthese* 198 (5): 4315–4330.

Tomalin, M. 2006. *Linguistics and the Formal Sciences.* Cambridge University Press.

Wilson, M. 2006. *Wandering Significance.* Oxford University Press.

12 The Scientist in the Machine

Rosa Cao

12.1 Introduction

Recent neural network models are impressive enough that calling them "artificial intelligence" has become commonplace. But their capabilities beg for explanation. What, for example, explains their linguistic fluency, and the surprisingly broad range of domains to which those language abilities sometimes appear to extend? And when they fail at tasks that are trivial for us (and seem like they ought to be trivial for them), what predicts those failures in turn?[1]

We are in the strange position of both being able to build the things and still wringing our hands about how little we understand about how they work. This is in contrast to the usual thought, which is that in order to build a thing, you have to know how it works—or Feynman's dictum, that you don't count as knowing how a thing works unless you can build it. Now we could give a simple engineering answer to how they work—for example, citing computational components such as attention heads and the matrix operations that they perform—but that does not quite address the demand to explain why something that performs these well-characterized low-level operations in just this arrangement can somehow do so much (and yet not enough). Why is it so hard to predict ahead of time which simple tasks they will continue to fail and which spectacular-seeming tasks the very same system will surprisingly master?

This question can also be asked in a more skeptical register. For example, we can ask whether, despite their linguistic fluency, language models really understand the words that they are producing, or whether predictively successful models of various phenomena are really explanatory, or merely curve-fitting.[2] Opinions differ wildly. Vast research programs proceed on the assumption that

1. Some examples of "surprising" failures for leading models as of early 2024 include long-form multiplication, playing tic-tac-toe, and other simple games (without explicitly calling for the model to write and run code).

2. See, for example, Bender and Koller's (2020) "stochastic parrots" paper, or Bowers et al. (2023) BBS target article and responses. For an opinionated overview on the sprawling debate, see Mitchell and Krakauer (2023).

folk psychology is directly and fruitfully applicable to LLMs, ascribing to them such capacities as "reasoning" or "thought."[3] Skeptics, by contrast, think this is all a terrible mistake, that we have somehow forgotten that these systems are mere tools; they do not have thoughts or feelings or goals, and however it is that they work, it is not in the same way that we do. Chomsky has dismissed them as a kind of "super auto-complete" (Goddu et al. 2024; Chomsky et al. 2023). This way of framing things naturally leads to a contrast: where if models are like us in the relevant respects, then they do really understand and use language in the same way that we do (and Searle was wrong). Whereas if models are not like us—if they work in a completely different way than we do—then they don't really understand what they are saying, and it's all just a bag of clever tricks.

How are we to adjudicate such disputes? One way is to look directly for similarities between models and brains. For example, if they are good models of us, then that may be because they resemble us in the relevant respects. Another is to get a better account of how they work at some illuminating level of description beyond the brute low-level computations that are performed, the architecture, the loss function, and the training data. We might intuitively think that it would help to pop the hood, and look "inside" to see whether there are internal representations that line up with our folk psychological interpretations—or whether there are internal representations at all. As a first pass, we might take the idea of "having a world model" or "understanding" as employing representations of the relevant portions of the world in order to succeed at a task in some domain.

Unfortunately, merely examining the internal workings of these systems will not yield anything like a dispositive answer. There is substantial disagreement on what it takes to be a representation and still more disagreement on how to map contents onto internal activations. Unless we know how to interpret internal activations, they will in a sense be "just more behavior." And so to the extent that looking at their behavior will be a guide to how these systems work and whether they "understand" what they are doing, internal activations will provide more helpful data—but they will not provide any sort of "ground truth." And that's before we even consider whether employing representations in some sense is enough for genuine understanding.

3. For just one strand in this work, see Wei et al. (2022) and the thousands of follow-up papers to the original "let's reason step-by-step" prompt. Other strands cite "deception" (Hagendorff 2024) or "beliefs" (Hase et al. 2023). Beyond AI researchers, philosophers and linguists have also embraced these possibilities: See, e.g., Capellen and Dever (2021), Lederman and Mahowald (2024).

That being said, it will still be useful to adopt a minimal notion of internal model, the presence or absence of which will capture some (but not all) of what is at stake in these questions. For present purposes, what it takes to have a minimal internal model is to have an internal representation of the relevant features of the task domain, and furthermore, for those internal representations to be causally implicated in successful task performance in that domain. Now what it takes to have a real internal representation in turn is also up for vociferous debate, but again, for present purposes, we can adopt a pragmatic, task-relative notion, where what is required is (1) that the information can be decoded from a system's internal physical states (subject to generally agreed-upon constraints about the sophistication and complexity of the decoder), and (2) that manipulations of that information lead to just the expected consequences of manipulating internal representations used for the task.[4]

Further bells and whistles (e.g., that the representation be structural, or that it must support the capacity for offline simulation to count as a genuine model, or that the information must be grounded in experience to count as a representation in the first place, or maybe just that there are other characteristics that we think any system capable of understanding must possess) can then be added to that requirement to get us to "real-world models" or "real representations" or "real understanding" for those who require more demanding notions to make the distinctions that they think are important in this broader debate about the capacities of generative AI systems.[5]

In what follows, I'll consider two related but conceptually distinct questions:

1. Are these models explanatory of some of the cognitive capacities that they mimic? (That is, are they good scientific models for us)?

2. Do these models have some of the cognitive capacities that they mimic? (That is, do they work the same way that we do—by modeling the world and understanding what they are doing?)

The first question arises in the context of NN models doing a good job both of mimicking human behavior at certain tasks, and (surprisingly) predicting

4. See Cao (2022) for further articulation of why these are plausible requirements for an explanatorily useful notion of representation. For examples of potential causal interventions, see Geiger et al. (2022) on interchange interventions, DeCao et al. (2021) on "knowledge editing" and Harding (forthcoming) for a summary of common techniques.

5. See, e.g., Yildirim and Paul (2024) on the "structure-preserving" requirement; Grush (1997), Ha and Schmidhuber (2018), and Diester et al. (2024) on the simulation requirement; and Goddu et al. (2024) on the requirement that a system must be an agent to count as having world knowledge. Cao and Warren (2022) spell out some implications of employing these different senses of "internal models." Contemporary discussions in turn have roots in an extensive older literature on the nature of cognitive maps, internal representations and models, going back to, e.g., Tolman (1948), Conant and Ashby (1970), Metzler and Shepard (1975), and Swoyer (1991).

brain activity in those same tasks. Surely being able to reproduce both behavior and the neural activity underlying it should count as a modeling success. And it might also lead us to think that the models are similar to their biological targets. But one might worry that for a model to explain a target, we need to know how the model itself works—something that wasn't a problem for more traditional physical or "hand-coded" models whose parts and settings had clear interpretations, but one that seems much more pressing for large many-parameter models constructed using deep learning. Hence the second question.

I'll argue in section 12.2 that success in predicting biological behavior is some (defeasible) evidence that these models are like us, in some respects, to some degree. section 12.3 will make the more general point that a positive answer to the explanatory question also points to a way of addressing the capacities question—namely, that one way of getting at how things really are is to do a good job of predicting the world by curve-fitting to enough data.[6] Perhaps you really can fake it until you make it.

Section 12.4 will examine some cases using the minimal notion of internal model proposed here, and then the concluding discussion in section 12.5 will explore different ways we might want to expand that minimal notion to accommodate more robust notions of representation or ideals of scientific understanding.

But the upshot will be that questions like: "Are these models for real, or are they somehow faking it?" do not admit of a clear-cut yes or no answer. Rather, there will be patterns in how the systems work, and how well those patterns map onto notions like "representing things" or "modeling the world" will be a matter of degree, and necessarily assessed relative to a particular context. And this will be just as true for us as it is for these artificial systems.

12.2 NN Models as Models of the Brain

Before tackling the more general question of how NN models succeed at various impressive tasks (when they do), let's look at the most obvious application of artificial neural networks—as models of biological neural networks. In the past ten years, task-optimized neural networks have been shown to be surprisingly good predictors of stimulus-evoked brain activity in visual, auditory, and language areas (Yamins et al. 2014, Kell et al. 2018, Schrimpf et al. 2021).

6. This seems to be an emerging theme in the more optimistic ML literature, from slogans like "Scale is all you need" (the glass half-full version of Sutton's Bitter Lesson), to Wei et al. (2022) on "emergent abilities," to more ambitious claims such as the "Platonic Representation Hypothesis" (Huh et al. 2024).

The worry is that with these and similar models, the ability to make good predictions of some target phenomenon is somehow not enough—a good model is supposed to do more than merely make good predictions if they are to be explanatory. (Compare the idea that producing accurate predictions is not always enough for acceptance of a theory as true. So too here, with the adjustment that because of the many ways that models are used, the explanatory goodness of a model need not depend on taking it to be "true"). To address the worry, we should either be able to say what further conditions must be met for explanatory goodness (and show that the models satisfy the conditions), or give an argument for why no further conditions are needed. But before that, it will be helpful to say a bit more about the nature of the predictive successes produced by these models.

Deep neural network models optimized for the task of image classification on ImageNet will exhibit patterns of activation at successive internal layers (conv1, conv2, conv3 . . .). Those activations can be fit to brain activity in corresponding regions of primate ventral visual stream (V1, V2, V4, IT), using a simple linear mapping. Using that same mapping (now fixed), experimenters can present novel images to both model and animal, and use the model activity to predict brain activity in these areas with higher accuracy than by any other means currently available. This is a striking result, since the competing models are ones where scientists have tried to specifically incorporate theories about what the brain is doing and brain data directly. For now, directly fitting the brain (in these particular tasks) does not yet work as well for predicting brain activity as fitting a proxy task for what the brain is capable of doing.[7]

What is interesting about these results is that the DNNs themselves have never been exposed to neural data—they were trained only to perform image classification, on a huge set of twenty million images that most monkeys have probably never been exposed to. Yet their activity maps to primate neural activity evoked by novel (held out) images by a simple linear recombination of units, yielding predictivity of near 80% accuracy for some areas and stimuli. Moreover, the most predictive DNN layers seem to correspond to brain areas in approximately the same relative location in the processing stream, so that

7. This is in part a contingent consequence of the relative paucity of brain data available today, compared to the plenitude of task training data. Monkey electrophysiology and human fMRI data are expensive; words on the internet (and even images and audio) are relatively cheap. When orders of magnitude more brain data becomes available (such as that now being collected by large collaborations), we should expect the direct fit to brain data to yield better predictions as well.

early layers correspond to primary sensory areas, and intermediate layers to intermediate sensory areas, and later layers to later sensory areas.[8]

12.3 Curve-Fitting as Model-Building

The rather particular success stories about NN models as predictively successful models of quite specific brain areas under certain task conditions illustrate a more general point, that one way to find real patterns in the world is by curve-fitting our data.

While the term is often used in a pejorative way—a model that is "merely" a curve-fit is supposed to be an uninformative post hoc summary of the data, one that offers no further insight or explanatory value—it need not be. Efficiently summarizing data is a way of finding regularities in the world, and could even help to identify the variables/entities involved. Hodgkin and Huxley famously found that the relationship between voltage and current across a cell membrane could be fit with a polynomial whose parameters ended up corresponding to the number of subunits in an ion channel. By literally curve-fitting the voltage/current relationship, they found that one of the parameters of that curve corresponded to a physical entity causally involved that relationship. Other examples abound: for example, Kepler's third law (about the relationship between the period of planetary orbits and their distance from the sun) constrained Newton's inverse-square law for gravity, or Faraday's law (eventually one of Maxwell's equations) from Faraday's empirical measurements of the relationship between the movements of a magnet and the voltage induced in a coil.

So curve-fitting can illuminate the variables involved in a phenomenon (although there are no guarantees). So far, this is unsurprising. But the brain example from the previous section also shows that curve-fitting can, at least sometimes, illuminate variables involved in generating very different data (from phenomena that are not obviously related at first glance). That is, it can

8. To be clear, the core model (task-optimized) has never been trained on brain data. The parameters of the linear mapping from model to brain are fit using neural data—but once trained, the linear mapping parameters are held fixed, and predictions are made from held-out model responses to held-out neural data. This is important, because one might worry that the linear mapping is itself encoding relevant brain data, in which case it would no longer be impressive that the task-optimized models were doing a good job of predicting brain data. But linear mappings alone (including those trained on all the brain data we have) are insufficient to perform the transformations on image inputs that are needed to generate good predictions of brain data—and this can be seen in controls where the linear predictivity of non-corresponding layers or ideal observer models is much lower than the linear predictivity of corresponding layers for the same brain area. See Schrimp et al. (2021) for details on one way these mappings are commonly performed, and for a current "leaderboard" on the most predictive models.

bring underlying connections to light, including ones not known when the measurements were being made. (Who would have predicted that the speed of light in a vacuum could fall out of measurements of electromagnetic phenomena in concrete objects?)

So far, I am making only a descriptive point: It is sometimes possible to predict one set of data by directly curve-fitting a different set of data. In the case of brain-prediction by curve-fitting image-to-label data, the result is striking. But in a weaker form, this happens every time we build a predictive model by fitting past data, and that model successfully predicts future data.

But there is an epistemic corollary: Getting a hard prediction right (one where the probability of success by chance is low) is some reason to think that the model is capturing a real pattern in the world, one responsible for generating the data being successfully predicted. This is because something like a *no miracles argument* can be applied, though in the context of models rather than theories.[9] When a model gets a prediction right, either it's successful just by chance, or because it's somehow tracking a real pattern in the world. If we have reason to think it's not just chance (perhaps because the probability of random success is very low, given the number of possibilities), then what remains is that we're capturing some aspect of the data generating process.[10]

So even if we reject strong similarities between artificial neural networks and biological brain targets, a more general point remains: that curve fitting can be illuminating when there is some systematic connection between the thing being fit and the thing being predicted. Perhaps they are similar (e.g., because the new data is drawn from the same distribution as the old), or perhaps there is some kind of systematic dependence relation between them, whether correlational, causal, or constitutive.

Given the examples above, we can see that curve-fitting might be expected to yield novel explanatory insights in just those cases where there are tight explanatory connections between the data being curve-fit and other related phenomena that were not directly measured. One famous example is the speed of light. Another might be the discovery of similar biological mechanisms in very different species, where "curve-fitting" what's happening in one species might tell you something about what's going on in the other (hence the utility of using model organisms). Those connections in turn could be causal connections due

9. In the spirit of Rowbottom et al. (2024).

10. This is a kind of realism, but far from industrial-strength. We might still be agnostic about whether the variables or parameters in a fit model correspond to real entities or fundamental quantities in the world (or which ones do and which don't, even in the same model).

to common ancestry, or quite different ones due to similar evolutionary pressures. But in both cases, there is some kind of illuminating dependency (either of phenotype on genes, or of phenotype on environmental pressures).[11]

This picture supports an account of explanation where what matters is getting robust dependencies right (as well as a sense of how robust they are, and what the boundaries of success and failure look like). So a model will be explanatory insofar as it captures real dependencies in the world. To the extent that curve-fitting sometimes allows us to do that, curve-fitting will also sometimes yield models that are (to that degree) explanatory as well as predictive. So what we might say about the vision models discussed earlier is that they are (at least!) partially explanatory models of a specific class of response properties to certain kinds of images in a selected set of brain regions—and they may also generalize to some extent beyond that—but they are far from complete models of primate visual capacities more generally.

Capturing real patterns is what allows for counterfactually robust predictions, and blurs the lines between "mere" curve-fits and explanatory models. We should expect the robustness of the patterns so found to vary, from the quite fragile ones that work only in very particular circumstances—"Where and when will Venus be observed in the night sky, from this geographical location?"—that we might call "phenomenological," to the mechanistic ones that license predictions in a wide range of counterfactual circumstances—"What would the orbit of Neptune be if Uranus didn't exist?"

Similarly, a mechanistic model in biology that breaks down a process into the working of various interacting parts to produce an outcome to be explained is one that allows us to make predictions about how the outcome depends on changes in the parts or their operations. But there is no deep distinction or sharp line between such a model and one that predicts merely that there is some relationship between pre-synaptic vesicle release and post-synaptic currents—only a difference in the richness and complexity of the pattern and the strength of the dependence between different parameters in the model.

If these claims are as reasonable as I'm trying to make them out to be, then what accounts for the dismissal of curve-fitting as nonexplanatory? In some cases, it seems that we can ascribe no physical interpretation to the parameters of the model. Unless the form of the parameterized model is suggestive of connections to physical entities, it can be hard to interpret parameters as corresponding to anything "real" in the world—anything that would license further inferences, broader predictions, manipulations, or connections to other things that we know. That makes it difficult to employ the models in a new

11. See Cao and Yamins (2024b) for a more detailed version of this argument.

context, because we don't know what parameters to tweak (unlike in the case of a mechanical model, where we know exactly what to change).

Just because it's not obvious from the surface form, however, does not mean that a model with a complex set of hard-to-interpret parameters is not empirically equivalent to a more elegant looking traditional "theory." (See Piantadosi [2023] for a series of illuminating examples in this vein, criticizing Chomsky's "Galilean" approach to theory-building in linguistics.)

In other cases, the models will capture a real pattern in the world, but fail when they go beyond the original data—the patterns might be weaker because they are more fragile. Consider an example of a theory that "fits the data," but which we think is inadequate: the Lamarckian model of the inheritance of acquired characteristics. On the one hand, it makes some true predictions: that offspring will tend to resemble their parents, and that the characteristics of organisms will tend to be well-suited to their success given the environments in which they need to survive. And these predictions are true (when they are), because they match past observations (of offspring resembling their parents, and adaptive characteristics). Occasionally, it is even more precisely accurate—as in cases of epigenetic inheritance, or when cultural inheritance plays a role in propagating learned information.

Why is the Lamarckian model often worse than the Darwinian model? One thought is that it makes fewer good predictions (it fits less data, and under fewer circumstances). And it makes many more false predictions (e.g., when acquired characteristics fail to be inherited). The blacksmith's child may have perfectly average-sized muscles in early childhood, even though the giraffe has a long neck at birth. But these are matters of degree, and broadness of applicability.

This might all seem very wishy-washy. One might be tempted to protest: But the Darwinian model is (more or less) true, and the Lamarckian one (basically) false, and not just because one makes many more true predictions and many less false ones than the other. I think that intuition comes from the further interpretation of these models as being committed to the existence of certain causal processes in the world, with constituents corresponding to particular entities (that may then fail to exist). Taken only as models that are committed to dependencies between this and that, but not the deep nature of the things involved in the dependencies, the differences in kind fade away. I'm not denying that these further commitments are a common part of scientific and potentially useful in their own right. But I think we can say useful things about explanation and models even setting those commitments aside.

As models become more robust and accurate, we might be more tempted to valorize them with the label of "explanatory" or "true." But instead of sorting models into "explanatory" and "not," it will be more useful to pay attention to how they vary along several different axes—from more generalizable and robust to less, from richer to poorer, better or worse for particular purposes. More precisely, I propose three main dimensions to focus on: (1) accuracy of predictions within the purported domain of application of the model, (2) scope or generalizability or robustness (i.e., the breadth of that domain, which we might also think of as "richness" in the sense of how many factors the model purports to account for the relations among), and (3) mappability or manipulability (that is, how well the parameters of the model map to real-world variables or entities that are at least in principle manipulable). We might also think of mappability as something like the interpretability of the model. These dimensions are illustrated in figure 12.1, along with a few salient examples. (Folk psychology is a fuzzy cloud somewhere in the middle, partly because it's not completely clear what it's committed to, how to measure its accuracy, or what physically manipulable variables its posits map onto).

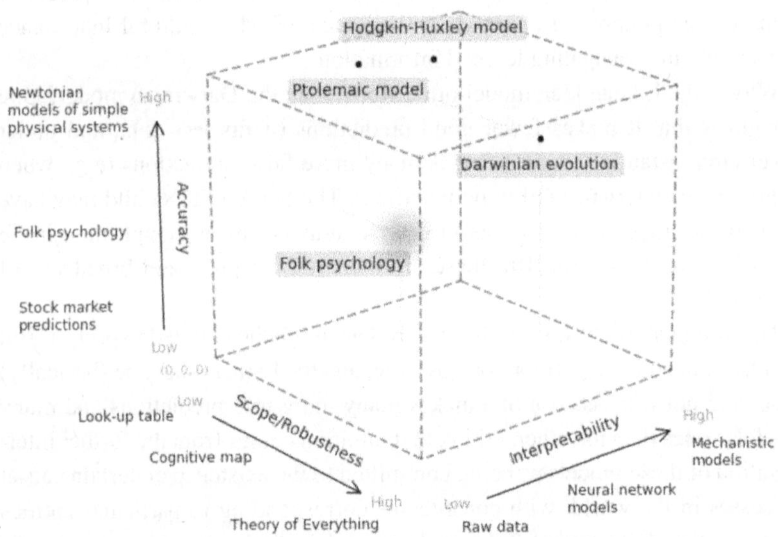

Figure 12.1: Dimensions of model explanatory value.

While these three dimensions are conceptually independent, there are systematic ways in which they could covary. As we extend the intended scope of a model (holding its actual features fixed), it will eventually generalize less well (and so will fall in accuracy). If we narrow the scope (e.g., to just the data to which it was originally fit), it will of course be more accurate. Similarly, if a model is not at all interpretable (and uses myriad parameters that don't correspond in any clear way to what seem to be relevant variables controlling variation in that domain), then we might expect it to be more fragile (lower in robustness and generalizability), and thus have narrower scope.

Of course, in the limit, we would have a model that made perfectly accurate predictions, was universally applicable, and was completely robust—and then we should be quite happy to call that the One True Theory of Everything.[12] In everyday life and everyday science however, our best models will be quite far from the limit, but may still be explanatorily useful.

12.4 Looking Inside the Black Box: What Does It Take to Have a World Model?

Having discussed scientific models and explanation, we can return to our motivating question, where we consider large neural networks trained to do certain tasks—that is, "generative AI models" such as LLMs—as potential modelers of the world in their own right (and not just as models that we employ as a guide to other phenomena). We might want to know things like: Are LLMs more like the rat with a cognitive map (perhaps no conscious plan, perhaps no intentional action, but with more flexibility in route selection than the dumb stimulus-response creature), or the man with a plan (conscious and intentional, perhaps self-reflective)? Or more like a tree that grows toward the light (with perhaps an implicit representation of where the light is), or a giant look-up table (with its traditional implication of having no model or representations at all)?

I suggested earlier that we can start with a common operationalization: having an "internal model" or "world model" of the relevant domain, understood in a relatively undemanding way as laid out in section 12.1. Then, in the spirit of the "dimensions" framework just introduced, we can see where a model's capacities fall along the dimensions of robustness, interpretability, and accuracy. As they increase along each dimension, I suspect we will find it more convincing to say that they have good internal models (and perhaps come closer to understanding, modulo the concerns to be addressed in section 12.5).

12. This way of looking at things is inspired by Mitchell (2000) as well as the conceptual spaces in Godfrey-Smith (2009).

Experiments show that it is possible to decode parts of speech, semantic similarity, grammatical structures, and so on from LLM activations using linear probes.[13] Looking outside the language domain, Li et al. (2023) showed that in a transformer trained on legal moves in the game Othello (and thus to predict future legal moves), they could not just decode board positions, but also manipulate them to change the output moves. These kinds of results have been taken by many researchers to show that models "represent" the world that their words describe, that they implicitly learn a model of the world from the data they are trained on.

This sort of evidence notwithstanding, it remains obvious to some people that LLMs do not really understand the stimuli that they respond to, or the words that they receive and produce. They are often cited as paradigm examples of how instrumental success can be achieved without understanding—unlike in our own case, where it is often presumed that we produce utterances because of what the words mean, and we intend to mean certain things by saying them.

It is equally obvious to others that they do have internal representations—ones that correspond to just those features that should figure in a theory of the domain in which they exhibit such surprising facility. How could they produce successful paraphrases without sensitivity to meaning? (Or more radically, what more is there to meaning than the ability to produce successful paraphrases?) How could they produce grammatical sentences as good as our own without at least the implicit grasp of grammar that humans have? (Nonlinguist humans, for the most part, do not have an explicit theory of grammar sufficient to generate their behavior). There are several observations to make here.

First, those who say that task success implies understanding, or having a world model, are making their own version of a no-miracles argument.[14] What could explain success at this very difficult task—so difficult we probably all thought it was impossible a decade ago—except that the model is somehow

13. See, for example, Li et al. (2021), Manning et al. (2020).

14. An extreme version of the view would say that competence is sufficient for (or even equivalent to having) internal representation. (See, e.g., Piantadosi and Gallistel 2024). And in light of the possibility of big look-up tables, this might seem obviously wrong. But of course, in the limit, having competence in all contexts (actual and counterfactual), robust to any perturbation, would be empirically indistinguishable from having understanding. So maybe the extreme version of the view is not as improbable as it might seem at first glance. But in our sublunary world of limited (even when impressive) competence, the best epistemic route to assessing understanding will be in terms of having competencies paired with the corresponding internal states posited by our best theory of the behavior. (Of course, this assessment will be limited by how good our theory of the behavior is as well.)

latching onto real structure in the domain? Or even in the far simpler Othello task—where the number of legal moves is still massively outmatched by the number of illegal ones? So whatever might make us doubt the no miracle argument in its original context should also make us doubt the claims here.

On the other hand, for the skeptics who think that because the models are "just" curve-fitting, they couldn't, even in principle, have a world model, we might point out that, just as with the scientific models discussed previously, sometimes by curve-fitting in one domain (word prediction), models can come to succeed in other domains that, while related, can also be perspicuously described in more semantically loaded ways (responding realistically to conversational prompts, summarizing long texts, and so on).[15] By forcing organisms to curve-fit to fitness, evolution produced some of the most sophisticated internal models in the world. So if models fail to have a world model, it must be for some other reason than just that they were created by curve-fitting.

And finally, just as it might be less helpful to talk about curve-fits vs. explanatory models, but rather dimensions of goodness for models, here too we may be better served by thinking in terms of a space of competence, in which models may exhibit sensitivities to real patterns in their task domain that vary along different axes (of accuracy, or scope and thus robustness, and of interpretability), corresponding to behavior that is more or less successful, more or less generalizable to new contexts, more or less robust to small changes.

Consider again a simple animal in a maze with a reward at the end. Suppose that the animal is capable of model-free learning only. It is sensitive to instrumental conditioning, and can be trained to make a series of correct turns. If there are multiple routes to the reward in the maze, it may be trained to take any of them. But if it sees the maze change so that a previously unavailable shortcut now becomes available, it will not be able to exploit that new knowledge, since (by supposition) it has no model of the overall spatial arrangement, nor how such an arrangement relates to getting reward. All it has is a stimulus-response pattern with no policy for novel stimuli.

Such an animal is less flexible than a more sophisticated one that is capable of model-based learning—of constructing a cognitive map that represents spatial structure in its environment, and using it to navigate. But there is also some sense in which the so-called model-free animal nonetheless does have a map—a very sparse one, supporting only brittle behavior—containing only the turns needed to get to its destination. And if we were able to examine all

15. More carefully: By curve-fitting in a few domains, including not just word prediction but instruction fine-tuning and maximizing reward for a model trained on human feedback, today's chatbots seem to have acquired a wide range of surprising competencies.

the activity in its brain as it navigates to reward, we could in principle decode some information about the local structure of the maze.

If we were to put the simple animal's internal model on the axes given in section 12.3, it might score quite high on accuracy (for that very route, barring any changes), but very low on scope/robustness (since brittle, can accommodate few changes in conditions), and also quite low on manipulability (since the behavioral policy has variables that map to animal-specific sensory responses to stimuli and animal-specific bodily actions that do not correspond in a straightforward way to geographical facts about the world (even though they do, with a sufficiently sophisticated decoder, carry some information about those facts)). The more sophisticated animal on the other hand would have a model with a higher scope/robustness score, and, depending on its internal architecture, perhaps a quite interpretable/manipulable one as well. That is, by modifying the right variables in the internal model, it might be possible to read off different consequences for potential routes.

In a recent study, Vafa et al. (2024) trained a 1.6 billion-parameter transformer model (with GPT-2 architecture) on turn-by-turn trajectories of taxi rides in Manhattan.[16] They found that their model was able to predict legal next-turns with 99% accuracy. More impressively, given enough routes (91 million) to curve-fit, the model can produce legal novel routes when given start and end-points. Is this enough to suppose that the transformer has extracted a map of Manhattan from the trajectories? The authors argue that it is not: because when trajectories are perturbed to require "detours" (by either randomly or adversarially changing the position mid-route), competence starts to break down—at 75% probability of an adversarial detour, their best model drops to 66% success at producing a valid route. (The invalid routes are shown in red in the inset, while valid routes are in black. We can see that the valid routes recapitulate much of the actual street map). By contrast, the "true model" that contains the actual street layout continues to succeed at 100% (the detours were chosen so that there was always some route to the destination still available).

The authors conclude that despite instrumental success at routing under normal (unperturbed conditions), their model does not have a "world model" (i.e., implicit map) of Manhattan. But if we examine the reconstructed map shown in of their paper (figure 12.2, what is striking is how much it does look like a map of Manhattan, albeit a very imperfect one. To be fair, some of this may

16. The authors' critique is that the models are incoherent and thus fragile—they fail to recognize that two sequences that result in the same state should accept the same continuations as valid, and/or they fail to recognize that there should be at least some differences in valid continuations for sequences that result in different states.

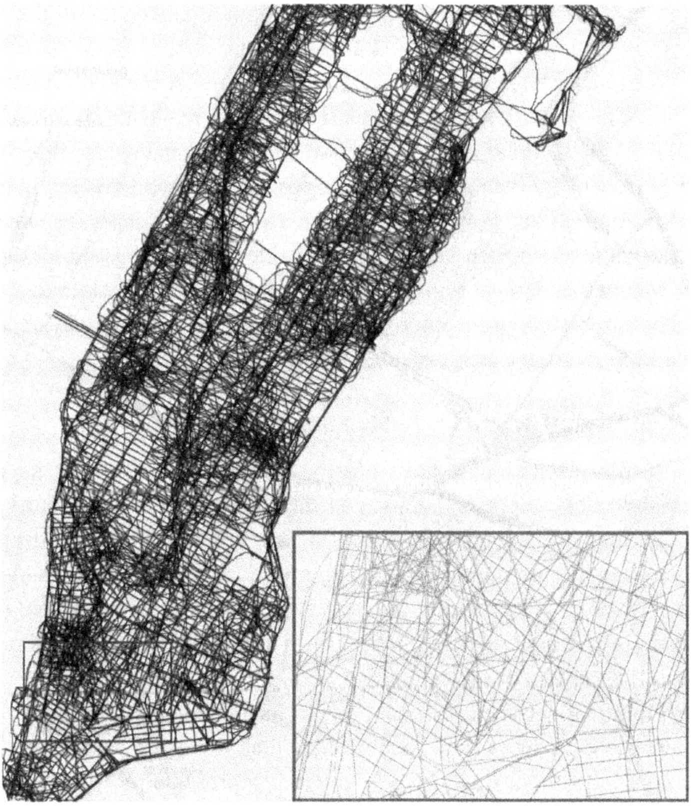

Figure 12.2: A reconstructed map of Manhattan shown in figure 3c from Vafa et al. (2024).

be in the eye of the beholder—we know what we are looking for, and there are hard constraints on the outer boundaries of the island, as well as some biases built into the reconstruction (vertices were given, but not connections among them). But what is left still captures real facts about the street geography that goes significantly beyond what was built in beforehand. We might say the map created by their best model has only medium accuracy, and low robustness (in the sense of only being applicable to a very specific domain and quickly useless if even small modifications of the task put it outside that domain), but still relatively high interpretability and mappability (since the authors were able to probe the "current intersection" based on turns made so far in a sequence, as well as make manipulations to the internal states to create detours).

I think the lesson to be drawn here is very much analogous to the one in the previous section: There is not an all-or-nothing contrast between having a world model and not. Rather, to have an internal model is to have competencies, more or less robust, that depend on sensitivity to real patterns in their data. Instead of trying to draw a bright line, we should instead assess the richness and robustness of the representations that we do find. While merely having a single procedure to get from point A to point B does not make for much of a map, being able to train on millions of taxi rides originating and terminating all over Manhattan can get you a more substantial one. And while that model is still far from perfect, it may not be so different from the ones carried around inside fallible creatures such as ourselves.

12.5 Representation and Understanding Revisited

So far in this discussion, I have mostly deferred issues about the nature of internal models, representations, and the more psychological attributions sometimes made of contemporary neural network models, suggesting only those who would employ more demanding notions may add requirements to the minimal sense here. In this last section, I'll look at some of these additional criteria and the role they might play in a more demanding (and perhaps more intuitive) notion of understanding.

In the minimal sense of section 12.4, to have a model of X is to exhibit competence with respect to X (competence that would be hard to explain—"miraculous"—without having some internal representation), to have some internal state that tracks the state of X (to be able to extract information about the state of X from the state of the system, using a probe that we think is legitimate), and to exhibit predictable effects on the system's competence when the content of that internal state is manipulated. To have a rich, or robust, or flexible model of X is to exhibit a broader array of competencies in a wider range of circumstances.

But a good old-fashioned AI (or GOFAI) chess-playing program would count as having genuine understanding of chess by these criteria. So would a GPS navigation system count as having a genuine understanding of the surrounding geography? If you think that verdict is appropriate, then the quick equivalence between understanding and satisfying these internal model criteria will be unproblematic. But if you think that understanding requires more, then LLMs will also not count as understanding.

What more is needed? Searle famously argued that competence cannot be enough for understanding. And although his argument is somewhat elliptical, his missing ingredient seemed to be something like phenomenology. We might think there is no understanding without the phenomenology of understanding (although this is contentious, and there can certainly be the phenomenology of understanding accompanying a failure of understanding as well). Or more directly, it might be that phenomenology—conscious phenomenology—is required for anything to count as a full-throated representation, of the kind whose possession counts as understanding.

With that view, we would need to answer the much more difficult question of whether these artificial systems are conscious (and when, and of what) first. I leave that task to others, since I think there are explanatorily useful notions of representation and understanding employed in science that do not require consciousness even if there are other notions that do.

An intermediate view also presents itself. One might think that there is a real contrast between our "unthinking" competence (at ball-catching or producing grammatical sentences in our native language) and our thinking competence (working through a mathematical argument or solving a design puzzle). The contrast between System 1 and System 2, for example, is one that often been raised in the context of discussion about LLMs and their competencies. If we want to associate "understanding" primarily with the explicit, slow, effortful representations manipulated by System 2 that are also associated with conscious thought, then that may be a functional distinction that is worth tracking. The greater functional integration with capacities in different domains may also support a greater range of the downstream cognitive consequences that we associate with understanding.

A related thought is that understanding is not just knowledge, and so acquiring more facts or storing more data is not enough. Rather, there are connections that have to be made between the relevant facts and what might be done with them, between the contingencies and dependencies that we become sensitive to and their relevance for future action. For the information encoded by an internal model to be integrated with information available in other modalities or subsystems, and then made broadly available to decision-making systems, we might need something like System 2. The relevance of action to understanding leads me to a second pressing worry, and one that is more often overlooked: the question of whether large neural network models have something like agency, and there it seems to me that there is more distance separating them from biological systems such as ourselves.

An example to illustrate the kind of worry I have in mind: Language models are notoriously bad at some simple games (24, pick the largest number, or tic-tac-toe), even though they are entirely capable of telling us the rules of these games (correctly), setting up environments to play these games, writing code to win at these games, and so on. But they do not seem to be able to play to win at these games, even when asked. It's as if they don't understand what winning is, even though they know how to use it in a sentence (rather like the psychopath, who is able to explain good and evil in perfect pedagogical terms, but is completely unmotivated to pursue one over the other). One possible explanation is that they might not want to win . . . because they don't want anything. Of course this is just one striking example, and other explanations are possible. But if they really don't have anything like an evaluative stance on the outcomes that they describe, then at least on some theories of meaning, that makes it impossible for them to count as understanding what winning—or any motivational term—means.

One response is that what language models have is still a partial understanding, because (on the view advocated in this chapter) their competence is narrow—limited to the context of talking, not to the context of putting words into action (even when those actions might be yet more text outputs). If their competence were broader, they would count as understanding after all. But I think this misses the point of the worry. Assessing the consequences of observations and how they matter for what I do next makes sense in the context of humans with goals and preferences in the world, wanting things to turn out one way rather than another. But it is much less clear that it makes sense in the context of artificial systems, which certainly have dispositions to produce certain outcomes over others. But do they have preferences or indeed any kind of evaluative stance toward outcomes, such that they will see some possibilities as being opened and others closed for potential action?

I said earlier that we could think about understanding as better or worse, corresponding to poorer or richer models that supported a narrower or broader range of competencies under different conditions. Having a broader competence, however, requires that when confronted with a new situation, you evaluate possible actions as positive or negative with respect to it, rather than having to be re-programmed to treat it as a task at all. So if we think of possibilities as restricted not just by what the system could do if it were re-oriented to try to address the new task (even without changing its internal "representations" learned previously), but by what it would do without explicit interference from us, it might turn out that for at least some artificial systems, there is no possibility of broader competence, even if the internal representations would, in principle, support them. In that case, one might think, again, that there is no understanding (and indeed, this might be what disqualifies the GPS system or the GOFAI chess system. The map information stored in the GPS system could be used (by some other agent) to figure out the total area of Central Park, but the system as configured is not able to deliver that answer. A person with an excellent internal map of Manhattan by contrast, could probably deliver a reasonable estimate).[17]

We might think that the Manhattan native in giving their estimate of the park's area will rely on their other capacities and knowledge (multiplication, addition, physical consequences of large vs. small areas), and perhaps that's part of the point as well: That understanding is at least partly holistic, and depends on grasping the connections between a target phenomenon and everything else that you know and can manipulate, and so any single-domain supercompetence will always be much more brittle and narrow than a messy human internal model that is nonetheless connected up with many other partial imperfect models of related aspects of the world. This is related to the earlier contrast between fast, automatic domain-specific competence and slower more effortful capacities that are characteristically more functionally integrated.

One last, perhaps more fundamental worry: Notions of representation that go beyond decodable information have a kind of normativity baked into them, to distinguish representation from misrepresentation, success from error, and that normativity depends on the consequences of the actions that the systems in question take being evaluated as successful or not. To the extent that that is the notion of representation underwriting our intuition that having an internal representation of the world is what it takes to understand the world (and succeed in tasks thereby), it may be that we should not ascribe representations to artificial systems in this way—or at least, not by default on the basis of probing experiments, or inherited from the way the term "representation" is used in machine learning. If the normativity comes only from our evaluation of these systems as succeeding or as failing, then any representational content we ascribe will ultimately depend on us as well. There is much more to say about where agency might come from (evolution, optimization), and how it might relate to evaluative attitudes and the normativity required for representation and misrepresentation, but I will leave that for other work.

Returning to our original question of how to understand LLMs and other neural network models, I think we should be open to seeing them as standing on a continuum with human scientists trying to understand the world. Sometimes we succeed through explicit understanding, creating elegant theories that capture broad swathes of natural phenomena. Other times, like limited models,

17. Perhaps this is an unfair comparison, since we might be implicitly provoking people's prosocial responses when we present them with a task (i.e., cueing them in the same way that a model might be provoked by a new prompt to do a new task). Of course, humans are constantly provoked to all kinds of responses and motivational attitudes by all kinds of environmental cues, perhaps quite subtle—whereas the models we have now are only provoked by a very limited set of inputs in a very limited set of formats. So perhaps this is only another (large) difference in degree, rather than in kind.

we curve-fit to local patterns but are unable to generalize to related domains. Yet other times, we curve fit, and add to our bag of tricks, and integrate the tricks with each other until our models become more robust, and practical success starts looking more and more like genuine understanding.

In that spirit, for us to understand how these systems work is just to identify robust patterns in their behavior, together with connections to the mechanistic descriptions of their internal workings—robust enough to support inference in future and wider contexts. Once we know how the systems work in this sense, disagreements about whether that counts as the systems themselves having genuine (or merely ersatz) understanding will be conceptual debates about how much we want to build into our notion of understanding, rather than empirical disagreements about the nature of these systems and how they differ (or don't) from ourselves.

References

Bender, E. M., and A. Koller. 2020. "Climbing Towards NLU: On Meaning, Form, and Understanding in the Age of Data." In *Proceedings of the Annual Meeting of the Association for Computational Linguistics,* 58:5185–5198.

Bowers, J. S., G. Malhotra, M. Dujmović, et al. 2023. "Deep Problems with Neural Network Models of Human Vision." *Behavioral and Brain Sciences* 46:e385. https://doi.org/10.1017/S0140525X22002813.

Cao, R., and J. Warren. 2023. "Mental Representation, "Standing-In-For", and Internal Models." *Philosophical Psychology,* 1–18.

Cao, R., and D. Yamins. 2024a. "Explanatory Models in Neuroscience, Part 1: Taking Mechanistic Abstraction Seriously." *Cognitive Systems Research* 85:101244.

Cao, R., and D. Yamins. 2024b. "Explanatory Models in Neuroscience, Part 2: Functional Intelligibility and the Contravariance Principle." *Cognitive Systems Research* 85:101200.

Cappelen, H., and J. Dever. 2021. *Making AI Intelligible: Philosophical Foundations.* Oxford University Press. https://doi.org/10.1093/oso/9780192894724.001.0001.

Chomsky, N., I. Roberts, and J. Watumull. 2023. *Guest Opinion Essay,* March.

Conant, R. C., and W. Ross Ashby. 1970. "Every Good Regulator of a System Must Be a Model of That System." *International Journal of Systems Science* 1 (2): 89–97.

De Cao, N., W. Aziz, and I. Titov. 2021. "Editing Factual Knowledge in Language Models." In *Proceedings of the 2021 Conference on Empirical Methods in Natural Language Processing,* 6491–6506. Association for Computational Linguistics.

Diester, I., M. Bartos, J. Bödecker, A. Kortylewski, C. Leibold, J. Letzkus, M. M. Nour, et al. 2024. "Internal World Models in Humans, Animals, and AI." *Neuron* 112 (14): 2265–2268.

Geiger, A., Z. Wu, H. Lu, J. Rozner, E. Kreiss, T. Icard, and C. Potts. 2022. "Inducing Causal Structure for Interpretable Neural Networks." In *International Conference on Machine Learning*, 7324–7338. PMLR.

Goddu, M., A. Noe, and E. Thompson. 2024. "LLMs Don't Know Anything: Reply to Yildirim and Paul." *Trends in Cognitive Sciences*, https://doi.org/10.1016/j.tics.2024.06.008.

Godfrey-Smith, P. 2009. *Darwinian Populations and Natural Selection*. Oxford University Press.

Grush, R. 1997. "The Architecture of Representation." *Philosophical Psychology* 10 (1): 5–23. https://doi.org/10.1080/09515089708573201.

Ha, D., and J. Schmidhuber. 2018. "World Models." *arXiv preprint*, arXiv: 1803.10122.

Hagendorff, T. 2024. "Deception Abilities Emerged in Large Language Models." *Proceedings of the National Academy of Sciences* 121 (24): e2317967121. https://doi.org/10.1073/pnas.2317967121.

Harding, J. Forthcoming. "Operationalising Representation in Natural Language Processing." *British Journal for the Philosophy of Science*.

Hase, P., M. Diab, A. Celikyilmaz, X. Li, Z. Kozareva, V. Stoyanov, M. Bansal, and S. Iyer. 2023. "Methods for Measuring, Updating, and Visualizing Factual Beliefs in Language Models." In *Proceedings of the 17th Conference of the European Chapter of the Association for Computational Linguistics*, 2714–2731. Association for Computational Linguistics.

Huh, M., B. Cheung, T. Wang, and P. Isola. 2024. "Position: The Platonic Representation Hypothesis." In *Proceedings of the 41st International Conference on Machine Learning*, 235:20617–20642. https://proceedings.mlr.press/v235/huh24a.html.

Kell, A. J., D. L. Yamins, E. N. Shook, S. V. Norman-Haignere, and J. H. McDermott. 2018. "A Task-Optimized Neural Network Replicates Human Auditory Behavior, Predicts Brain Responses, and Reveals a Cortical Processing Hierarchy." *Neuron* 98 (3): 630–644.

Lederman, H., and K. Mahowald. 2024. "Are Language Models More Like Libraries or Like Librarians? Bibliotechnism, the Novel Reference Problem, and the Attitudes of LLMs." *Transactions of the Association for Computational Linguistics* 12:1087–1103.

Li, B. Z., M. Nye, and J. Andreas. 2021. "Implicit Representations of Meaning in Neural Language Models." In *Proceedings of the 59th Annual Meeting of the Association for Computational Linguistics and the 11th International Joint Conference on Natural Language Processing (Volume 1: Long Papers)*, 1813–1827.

Li, K., A. K. Hopkins, D. Bau, F. Viégas, H. Pfister, and M. Wattenberg. 2023. "Emergent World Representations: Exploring a Sequence Model Trained on a Synthetic Task." In *ICLR*.

Manning, C. D., K. Clark, J. Hewitt, U. Khandelwal, and O. Levy. 2020. "Emergent Linguistic Structure in Artificial Neural Networks Trained by Self-Supervision." *Proceedings of the National Academy of Sciences* 117 (48): 30046–30054.

McNamee, D., and D. M. Wolpert. 2019. "Internal Models in Biological Control." *Annual Review of Control, Robotics, and Autonomous Systems* 2 (1): 339–364. https://doi.org/10.1146/annurev-control-060117-105206.

Metzler, J., and R. N. Shepard. 1974. "Transformational Studies of the Internal Representation of Three-Dimensional Objects." In *Theories in Cognitive Psychology,* edited by R. L. Solso, 147–201. Routledge.

Mitchell, M., and D. C. Krakauer. 2023. "The Debate over Understanding in AI's Large Language Models." *Proceedings of the National Academy of Sciences* 120 (13): e2215907120.

Piantadosi, S. T. 2023. "Modern Language Models Refute Chomsky's Approach to Language." In *From Fieldwork to Linguistic Theory: A Tribute to Dan Everett,* edited by E. Gibson and M. Poliak, 353–414.

Piantadosi, S. T., and C. R. Gallistel. 2024. "Formalising the Role of Behaviour in Neuroscience." *European Journal of Neuroscience* 60 (5): 4756–4770. https://doi.org/10.1111/ejn.16372.

Rowbottom, D. P., W. Peden, and A. Curtis-Trudel. 2024. "Does the No Miracles Argument Apply to AI?" *Synthese* 203 (5): 1–20.

Schrimpf, M., I. A. Blank, G. Tuckute, C. Kauf, E. A. Hosseini, N. Kanwisher, et al. 2021. "The Neural Architecture of Language: Integrative Modeling Converges on Predictive Processing." *Proceedings of the National Academy of Sciences* 118 (45): e2105646118.

Schrimpf, M., J. Kubilius, M. J. Lee, N. A. Ratan Murty, R. Ajemian, and J. J. DiCarlo. 2020. "Integrative Benchmarking to Advance Neurally Mechanistic Models of Human Intelligence." *Neuron* 108 (3): 413–423. https://www.sciencedirect.com/science/article/pii/S089662732030605X.

Swoyer, C. 1991. "Structural Representation and Surrogative Reasoning." *Synthese* 87:449–508.

Tolman, E. C. 1948. "Cognitive Maps in Rats and Men." *Psychological Review* 55 (4): 189.

Tuckute, G., J. Feather, D. Boebinger, and J. H. McDermott. 2023. "Many but Not All Deep Neural Network Audio Models Capture Brain Responses and Exhibit Correspondence Between Model Stages and Brain Regions." *PLOS Biology* 21 (12): e3002366.

Vafa, K., J. Y. Chen, J. Kleinberg, S. Mullainathan, and A. Rambachan. 2024. "Evaluating the World Model Implicit in a Generative Model." *arXiv preprint,* eprint: arXiv: 2406.03689.

Wei, J., Y. Tay, R. Bommasani, C. Raffel, B. Zoph, S. Borgeaud, and W. Fedus. 2022. "Emergent Abilities of Large Language Models." *arXiv preprint,* arXiv: 2206.07682.

Wei, J., X. Wang, D. Schuurmans, M. Bosma, B. Ichter, F. Xia, E. H. Chi, Q. V. Le, and D. Zhou. 2022. "Chain-of-Thought Prompting Elicits Reasoning in Large Language Models." In *Proceedings of the 36th Advances in Neural Information Processing Systems (NeurIPS),* 24824–24837.

Yamins, D., H. Hong, C. F. Cadieu, E. A. Solomon, D. Seibert, and J. J. DiCarlo. 2014. "Performance-Optimized Hierarchical Models Predict Neural Responses in Higher Visual Cortex." *Proceedings of the National Academy of Sciences* 111 (23): 8619–8624.

Yildirim, I., and L. A. Paul. 2024. "From Task Structures to World Models: What Do LLMs Know?" *Trends in Cognitive Sciences* 28 (5): 404–415.

Contributors

Aleksandr Alekseev

Aleksandr Alekseev is an assistant professor at the Department of Economics at the University of Regensburg (Germany). His primary research field is behavioral labor economics. He studies the interaction between the current trends in the labor market and human psychology. He is interested in the behavioral effects of automation in the workplace and how they shape workers' welfare and inequality. His other research topics include the role of ability and motivation in success in life, the determinants of effort, and the methodology of experimental economics.

Rosa Cao

Rosa Cao is an assistant professor of philosophy at Stanford University. Cao received her PhD in brain and cognitive sciences from MIT in 2010 and her PhD in philosophy from New York University in 2018. Her research is focused on what might be broadly termed "informational" approaches to studying the brain, mind, and cognition, and addresses such questions as: How should we understand notions such as "representation," "computation," and "information" in neuroscientific and philosophical frameworks? How are computational models to be evaluated empirically? What is the status of neural representation, and how does it relate to mental representation? How should we understand the relationship between natural and artificial intelligence?

Sean M. Carroll

Sean Carroll is Homewood Professor of Natural Philosophy at Johns Hopkins University, and fractal faculty at the Santa Fe Institute. He received his PhD in 1993 from Harvard University. His research focuses on foundational questions in quantum mechanics, spacetime, cosmology, emergence, entropy, and complexity, occasionally touching on issues of dark matter, dark energy, symmetry, and the origin of the universe. Carroll is the author of *The Biggest Ideas in the Universe*, *Something Deeply Hidden*, *The Big Picture*, *The Particle at the End of the Universe*, *From Eternity to Here*, and *Spacetime and Geometry: An Introduction to General Relativity*. He has been awarded prizes and fellowships by the National Science Foundation, NASA, the Sloan Foundation, the Packard Foundation, the American Physical Society, the American Institute of Physics, the American Association for the Advancement of Science, the Freedom From Religion Foundation, the Royal Society of London, and the Guggenheim Foundation. Carroll has appeared on TV shows such as The *Colbert Report*, PBS's *NOVA*, and *Through the Wormhole* with Morgan Freeman, and frequently serves as a science consultant for film and television. He is host of the weekly *Mindscape* podcast. He lives in Baltimore with his wife, writer Jennifer Ouellette.

Glenn W. Harrison

Glenn Harrison is a distinguished university professor, the C. V. Starr Chair of Risk Management & Insurance, and director of the Center for the Economic Analysis of Risk in the Maurice R. Greenberg School of Risk Science, J. Mack Robinson College of Business, Georgia State University, Atlanta. His current research interests include risk preferences, risk perceptions, risk management, behavioral welfare economics, Bayesian econometrics, and experimental economics. Google Scholar records that his published research has over 31,000 citations and a h-index of 81. Before academic life, Professor Harrison played Australian "no-rules" football for Hawthorn in the Australian Football League.

Margarida Hermida

Margarida Hermida is the current Jacobsen Fellow in Philosophy at King's College London. Her postdoctoral research project is on "The Physics of Life," and her aim to study how the philosophy of biophysics can shed light on the relationship between physics and biology, and on the nature of life in the universe. She received her PhD in philosophy from the University of Bristol, supervised by Samir Okasha, for which she was awarded a scholarship by the British Society for the Philosophy of Science (BSPS). Her thesis on "The Life and Death of Animals" developed a new metaphysical view of the nature of organisms and how they persist over time. She also holds a PhD in biology from the University of Porto (Portugal), and previously worked as a marine biologist.

Harold Kincaid

Harold Kincaid is emeritus professor and senior scholar in the School of Economics, University of Cape Town, South Africa. His work is in the philosophy of science, especially the philosophy of the social and behavioral sciences. Kincaid is the author, editor, or coeditor of sixteen books in the field and numerous journal articles and book chapters. He is currently pursuing issues in causal modeling and in experimental economics work on risk and time attitudes, trust, and beliefs.

James Ladyman

James Ladyman is professor of philosophy at the University of Bristol and President of the British Society for the Philosophy of Science 2023–2025. He has published many articles largely in philosophy of science, and several books including *Understanding Philosophy of Science*, (with Don Ross) *Every Thing Must Go: Metaphysics Naturalised*, (with Robin Brown) *Materialism: A Historical and Philosophical Introduction*, and (with Karoline Weisner) *What Is a Complex System*? He founded the University of Bristol's Centre for Science and Philosophy. His current work focuses on the relationship between physics and chemistry and biology.

Morten Lau

Morten Lau is professor of experimental economics at Copenhagen Business School, Denmark, professor of Behavioral Finance at Durham University, United Kingdom, and honorary professor at the School of Economics, University of Cape Town, South Africa. He has published in top journals in economics, such as *American Economic Review* and *Econometrica*. His research is mainly focused on decision making under risk and uncertainty and choice over time delay. Currently his main research interest is elicitation and estimation of higher order risk attitudes.

James W. McAllister

James W. McAllister is professor of history and philosophy of science at the University of Leiden and a member of the Leiden Institute for Brain and Cognition. He is the author of *Beauty and Revolution in Science* (Cornell University Press, 1996). As well as the relation between patterns in empirical data and structures in the world, his research focuses on the roles of common sense in the natural sciences and *Geisteswissenschaften* and connections between philosophy and history of science. He is *membre titulaire* of the Académie Internationale de Philosophie des Sciences, Brussels.

Tyler Millhouse

Tyler Millhouse is an assistant professor of practice in the University of Arizona's College of Information Science. He previously served as a postdoctoral fellow at the Santa Fe Institute, working on the "Foundations of Intelligence in Natural and Artificial Systems" project. Millhouse's research seeks to understand how concepts from AI and machine learning can inform the philosophy of cognitive science and the philosophy of science more generally. His work has appeared in leading journals, such as *The Australasian Journal of Philosophy*, *The British Journal for the Philosophy of Science*, and *Philosophy of Science*. He received his PhD in philosophy from the University of Arizona in 2021.

Ryan M. Nefdt

Ryan M. Nefdt is professor of philosophy at the University of Cape Town and a senior fellow in philosophy at the University of Bristol. He works broadly within cognitive science with a specialization in the philosophy of linguistics. He has produced numerous articles and book chapters on the philosophies of linguistics, science, AI, and language in various journals such as *Philosophy of Science*, *Biology & Philosophy*, *Mind & Language*, *Linguistics & Philosophy*, the *Journal of Linguistics*, and more. He has also written a number of books, including *Language, Science, and Structure* (OUP 2023), The *Philosophy of Theoretical Linguistics* (CUP 2024), and *Linguistic Relativity* (OUP forthcoming, with Jeffry Pelletier).

Achyuth Parola

Achyuth Parola is currently an undergraduate at Johns Hopkins University studying biophysics, applied mathematics and statistics, and philosophy. He is a recipient of the Woodrow Wilson Fellowship for undergraduate research in biophysics and philosophy. His current main research projects concern the understanding of complexity and emergence within the philosophy of science (with Sean Carroll), the development of ethics for probabilistic decision-making within public health policy, and the usage of reaction-diffusion models to understand receptor and ligand trafficking within cells.

Steve Petersen

Steve Petersen is Professor of Philosophy at Niagara University. His work is on computational approaches to "good thinking," broadly construed. More narrowly speaking, this research has two main tracks: applying algorithmic information theory to foundational issues in philosophy of science, and seeking a general theory of agency that would be informative about future AI values. In both he has drawn crucial inspiration from Daniel Dennett's work. He has been supported by the Survival and Flourishing Fund, the Center for Effective Altruism's Long-Term Future Fund, the Future of Life Institute, Nonlinear, Ramana Kumar, and the Center for AI Safety.

Don Ross

Don Ross is professor and head of the School of Society, Politics, and Ethics at University College Cork, Ireland; professor in the School of Economics at the University of Cape Town, South Africa; and program director for Methodology at the Center for the Economic Analysis of Risk, Robinson College of Business, Georgia State University, Atlanta. He is the author or editor of eighteen books, most recently *The Gambling Animal* (with Glenn Harrison) and numerous articles. His current main research projects concern risky choice experiments with African elephants, the economics of elephant conservation, and mathematical foundations to unify economic and sociological models of information and influence propagation in social networks (with Wynn Stirling)

David Wallace

David Wallace is the A. W. Mellon Professor of Philosophy at the University of Pittsburgh, and works on the interface between physics and philosophy. His research interests span the philosophy of contemporary physics, including philosophy of spacetime and symmetry, quantum field theory and quantum gravity, statistical mechanics, and the metaphysics of physics, especially concerning emergence and structural realism. He is especially known for his work developing and defending the Everett interpretation of quantum mechanics (aka the "many-worlds theory"), which he has defended as the most conservative and naturalistic way to approach quantum theory despite its apparently science-fictional nature.

Publisher contact:
The MIT Press
Massachusetts Institute of Technology
77 Massachusetts Avenue, Cambridge, MA 02139
mitpress.mit.edu

EU Authorised Representative:
Easy Access System Europe, Mustamäe tee 50,
10621 Tallinn, Estonia
gpsr.requests@easproject.com

Printed by Integrated Books International,
United States of America